AA004506

MINOS 2nd International Workshop, Irradiation of Nuclear Materials

Flux and Dose Effects

EPJ Web of Conferences Volume 115 (2016)

Cadarache, France
4 - 6 November 2015

Editor:

Christophe Galle

ISBN: 978-1-5108-2209-2

Printed from e-media with permission by:

Curran Associates, Inc.
57 Morehouse Lane
Red Hook, NY 12571

Some format issues inherent in the e-media version may also appear in this print version.

Printed by Curran Associates, Inc. (2016)

For additional information, please contact EDP Sciences – Web of Conferences at the address below.

EDP Sciences – Web of Conferences
17, Avenue du Hoggar
Parc d'Activité de Courtabœuf
BP 112
F-91944 Les Ulis Cedex A
France

Phone: +33 (0) 1 69 18 75 75
Fax: +33 (0) 1 69 28 84 91

contact-edps@webofconferences.org

Additional copies of this publication are available from:

Curran Associates, Inc.
57 Morehouse Lane
Red Hook, NY 12571 USA
Phone: 845-758-0400
Fax: 845-758-2633
Email: curran@proceedings.com
Web: www.proceedings.com

2nd MINOS workshop: program

Wednesday, November 4, 2015

8:15	•	Registration
9:00	•	Welcome, **C. Gallé (MINOS)**, CEA Saclay (France)
9:05	•	Opening Talk, **J.-P. Bouyer (DEN/DANS/Dir)**, CEA Saclay (France)

SESSION 1: IRRADIATION TOOLS AND FACILITIES, NOV. 4TH, (a.m.)

Chairman: R. P. Webb, University of Surrey (UK) – Co-Chairman: J.-L. Béchade, CEA Saclay (France)

9:20	•	**EMIR Federation: Network of Accelerators Dedicated to Material Irradiation, S. Bouffard et al., CEA Saclay (France)** pg. 1
10:00	•	Surrey Ion Beam Centre: the EPSRC "MRF" for Ion Beam Applications, **R. P. Webb**, University of Surrey (Surrey, UK) pg. 3
10:25	•	The Jules Horowitz Reactor Research Project : a New High Performance Material Testing Reactor Working as an International User Facility – First Developments to Address R&D on Material, **G. Bignan et al.**, CEA Cadarache (France) pg. 5
10:50		Coffee Break
11:15	•	The Characteristics and Irradiation Capabilities of MARIA Research Reactor in NCBJ Świerk, **G. Krzysztoszek**, NCBJ (Świerk, Poland) pg. 7
11:40	•	Plasma Material Interaction in Tokamak: the Contribution of WEST and of Laboratory studies, **C. Grisolia et al.**, CEA Cadarache (France) pg. 9
12:05	•	Discussion
12:35		Lunch Break

SESSION 2: PRIMARY DAMAGE AND STRUCTURAL DEFECTS, NOV. 4TH, (p.m.)

Chairman: F. Gao, University of Michigan (USA) – Co-Chairman: J.-P. Crocombette, CEA Saclay (France)

14:00	•	**Modeling and Simulation of Primary Damage and Structure Evolution in Ceramics and Metals, F. Gao, University of Michigan (Ann Arbor, USA) pg. 11**
14:40	•	DART, a BCA Code to Assess and Compare Primary Irradiation Damage in Nuclear Materials Submitted to Neutron and Ion Flux, **L. Lunéville et al.**, CEA Saclay (France) pg. 13
15:05	•	Polymers under Ionizing Radiations: the Specificity of Swift Heavy Ions, **Y. Ngono-Ravache et al.**, CEA Saclay (France) pg. 15
15:30		Coffee Break
15:55	•	Structure of Defects and Microstructure Evolution in Oxide Ceramics – Role of Electronic Excitation and Selective Displacement Damage, **K. Yasuda**, Kyushu University (Fukuoka, Japan) pg. 17
16:20	•	Irradiation Induced Phenomena in Nuclear Glass, **J-M. Delaye et al.**, CEA Marcoule (France) pg. 19
16:45	•	Simulation of C-Component Dislocation Loops by Ions and Protons Irradiation. A Tool to Understand the Breakaway Growth of Recrystallized Zr-Based Alloys, **J-P. Mardon et al.**, AREVA NP (Lyon, France) pg. 21
17:10		Discussion
17:40		End of the First Day

18:30		Welcome Cocktail (CEA Cadarache guest house, 18h30 – 20h30)

2nd MINOS workshop program (continued)

Thursday, November 5, 2015

SESSION 3: EVOLUTION & CHARACTERIZATION OF THE MICROSTRUCTURE, NOV. 5TH, (a.m.)

Chairman: M. J. Caturla, University of Alicante (Spain) – Co-Chairman: G. Jomard, CEA Cadarache (France)

9:00	•	**Microstructure Evolution in Fe and Fe-Cr Alloys with OKMC Methods, M. J. Caturla et al., University of Alicante (Alicante, Spain) pg. 23**
9:40	•	Atomistic Modeling of Segregation and Precipitation in Fe-Cr Alloys under Irradiation, **F. Soisson et al.**, CEA Saclay (France) pg. 25
10:05	•	Study of Rare Gases Behavior in Uranium Dioxide, **G. Carlot et al.**, CEA Cadarache (France) pg. 27
10:30		Coffee Break
10:55	•	Positron Annihilation Spectroscopy to Characterize Irradiation Induced Vacancy Type Defects in Materials for Nuclear Fission & Fusion, **M-F. Barthe**, CNRS-CEMHTI (Orléans, France) pg. 29
11:20	•	α Self-Irradiation Effects on Structural Properties of (U,Am)$O_{2\pm\delta}$ Materials, **F. Lebreton et al.**, CEA Marcoule (France) pg. 31
11:45	•	Discussion
12:20		Lunch Break

SESSION 4: IMPACT ON PHYSICAL, CHEMICAL, MECHANICAL CHARACTERISTICS & PROPERTIES, NOV. 5TH, (p.m.)

Chairman: I. M. Robertson, University of Wisconsin-Madison – Co-Chairman: D. Féron, CEA Saclay (France)

14:00	•	**Dislocation Grain Boundary Interactions in Irradiated Metals, I. M. Robertson et al., University of Wisconsin-Madison (Madison, USA)** pg. 33
14:40	•	Effects of Clear Bands on Intergranular Stresses and IASCC Early Damage, **M. Sauzay et al.**, CEA Saclay (France) pg. 35
15:05	•	Evolution under Irradiation of Optimized Austenitic Steel for Gen-IV Reactors, **A. Courcelle et al.**, CEA Saclay (France) pg. 37
15:30	•	Multiscale Modeling of Radiation Hardening in RPV and Austenitic Stainless steels: from the Atomic to the Continuum Scale, **G. Monnet**, EDF R&D (Moret-sur-Loing, France) pg. 39
15:55		Coffee Break
16:20	•	High Burn-up Structure in Nuclear Fuel: Impact on Fuel Behavior, **J. Noirot et al.**, CEA Cadarache, (France) pg. 41
16:45	•	Influence of Irradiation on Stainless Steel Corrosion in PWR Primary Conditions, **S. Perrin et al.**, CEA Saclay, (France) pg. 43
17:10	•	Discussion
17:40		Break
17:50	•	Closing Statements, **F. Carré (DEN/DS)**, CEA Saclay (France)

18:00		End of the Second Day

2nd MINOS workshop program (continued)

Friday, November 6, 2015

JULES HOROWITZ REACTOR AND LECA-STAR LABORATORY VISITS*, NOV. 6TH, (a.m.)

(*): limited participation

8:00	•	Meeting at the guest house parking and departure (by bus) to the main gate of CEA Cadarache for entrance procedure
8:45	•	First group to visit RJH reactor and LECA-STAR facility
10:45	•	Transfer
11:00	•	Second group to visit RJH reactor and LECA-STAR facility
13:00	•	End of visits, back to the guest house
13:15		Lunch Break (guest house)
14:00	•	End of MINOS Workshop

Keynote and lecture speakers

- Serge Bouffard, CEA Saclay (France)
- Maria José Caturla, University of Alicante (Spain)
- Fei Gao, University of Michigan (USA)
- Ian M. Robertson, University of Wisconsin-Madison (USA)

- Marie-France Barthe, CNRS-CEMHTI (France)
- Gilles Bignan, CEA Cadarache, JHR (France)
- Jean-Marc Delaye, CEA Marcoule (France)
- Gaelle Carlot, CEA Cadarache (France)
- Arnaud Courcelle, CEA Saclay (France)
- Christian Grisolia, CEA Cadarache (France)
- Grzegorz Krzysztoszek, NCBJ (Poland)
- Florent Lebreton, CEA Marcoule (France)
- Laurence Lunéville, CEA Saclay (France)
- Jean-Paul Mardon, AREVA NP (France)
- Ghiath Monnet, EDF R&D (France)
- Yvette Ngono, CEA Saclay (France)
- Jean Noirot, CEA Cadarache (France)
- Stéphane Perrin, CEA Saclay (France)
- Maxime Sauzay, (CEA Saclay (France)
- Frédéric Soisson, CEA Saclay (France)
- Roger P. Webb, University of Surrey (UK)
- Kazuhiro Yasuda, Kyushu University (Japan)

Local scientific committee members

- Jean-Luc Béchade, CEA Saclay (France)
- Lucile Beck, CEA Saclay (France)
- Philippe Bossis, CEA Saclay (France)
- Damien Féron, CEA Saclay (France)
- Muriel Ferry, CEA Saclay (France)
- Christian Grisolia, CEA Cadarache (France)
- Gérald Jomard, CEA Cadarache (France)
- Bernard Marini, CEA Saclay (France)
- Maylise Nastar, CEA Saclay (France)
- Jean Noirot, CEA Cadarache (France)
- Sylvain Peuget, CEA Marcoule (France)
- Sylvie Pillon, CEA Marcoule (France)
- Benoît Tanguy, CEA Saclay (France)
- Carole Valot, CEA Cadarache (France)

Organization

- Christophe Gallé, CEA/MINOS and Corine Farnaud-Pasquier, CEA/Cadarache
- Constantin Meis and Mireille Cestrières, CEA/INSTN, Saclay

Sponsors

MATERIALS

AREVA

Materials Simulation & Engineering

Réacteur Jules Horowitz Reactor

Scope and objective of the workshop

Created in 2011 by the Nuclear Energy Division of CEA, the French leader for research on materials for nuclear applications, the **MINOS Centre of Excellence for Nuclear Materials** sustains fundamental and applied research related to the behavior of nuclear materials and structures in severe environments. The main objectives of MINOS research are increasing the safety level and lifespan of current reactors, optimizing waste management, and developing new materials for generation IV systems, in particular fast neutron reactors. To accomplish these goals, it is essential to identify and understand the mechanisms of irradiation damage and to develop predictive models. These models are not only necessary to describe irradiation-induced microstructure evolution but also to predict the role of this microstructure in the evolution of the mechanical behavior and the material properties. In this domain, neutron irradiation is an important source of atomic-scale microstructural disorders (cascades) that generate a large number of structural defects (vacancies, loops, solute clusters, segregation, precipitates…). These defects could lead to various phenomena that affect the macroscopic physical and mechanical properties, such as hardening, embrittlement, swelling, creep, and cracking. These detrimental phenomena can reduce the efficiency and operability of nuclear power plant parts and components (i.e. fuel, claddings, internals, pressure vessels for PWR's).

MINOS intends to promote research on nuclear materials, to reinforce the visibility of CEA in this field, to identify new topics of research, and to contribute to training activities in materials science. To help achieve these objectives, MINOS organized an international workshop on "Materials Innovation for Nuclear Optimized Systems" in December 2012 and an international school on the "Modeling of Corrosion for Nuclear Reactors" in April 2014.

The 2015 MINOS international workshop on "Irradiation of Nuclear Material: Flux and Dose Effects" will provide the opportunity to address specific topics related to state-of-the-art research on the effects of irradiation on materials used in existing reactors and materials intended for future nuclear systems for fission or fusion. A large part of the workshop will be dedicated to structural materials (mainly metallic alloys) and fuels materials, but other materials, such as composites, nuclear glass, and polymers, will also be considered. Emphasis will be given to flux and dose effects. The workshop will consist of four sessions, and each one will combine keynotes by leading scientists and lectures by experts in materials science and nuclear engineering.

The workshop will be held at the National Institute for Nuclear Science & Technology (INSTN) in Cadarache, France. It is organized by the CEA Nuclear Energy Division with the support of the INSTN.

Main comments and outcomes of session 1: Irradiation Tools & Facilities

Chair: R. P. WEBB, Surrey Ion Beam Centre, University of Surrey (Guilford, UK)
Co-chair: J.-L. BÉCHADE, CEA-DEN-DMN, Service de Recherches de Métallurgie Physique, SRMP (Saclay, France)

From the presentation of S. Bouffard, it arises that thanks to the EMIR platform (six facilities), a large range of irradiation conditions are available in France. The Surrey Ion Beam Centre (R. P. Webb talk) provides complementary facilities mainly devoted to support micro-electronics irradiation damage studies for industry (with an intensive quality insurance and standards program). On the other side, it was underlined that Materials Testing Reactors are also able to provide useful complementary data about irradiation effects with neutrons: MARIA reactor in Poland (NCBJ) till 2030 (G. Krzysztoszek presentation) and RJH (C. Colin and G. Bignan presentation) which is under construction in CEA Cadarache and should be in operation at the end of 2020. For RJH, impressive devices will be proposed to study in service but also incidental and accidental conditions for RPV, internals and claddings of PWR. Finally a good illustration of the need to use modelling approaches and experimental irradiation was illustrated by the west project (C. Grisolia) where tungsten-based material is submitted to high thermal gradient and high flux of hydrogen isotopes.

Finally, main concerns shared by the participants have shown that:
- Access to charged particles irradiation facilities could be complex, Ease of access to European Facilities could be improved. The mixed funding model, even within CEA for example, causes some confusion - some provide free (at the point of use) access, some cost real money. There are also scheduling issues for some when there is only a single call for users each year.;
- It seems that facilities in Europe are enough for what research teams want to investigate although there are higher energy facilities in Japan and USA but it is essential to keep and develop irradiation tools and to reinforce the complementarity between accelerators and research reactors ;
- The question of the representativeness of neutron irradiation using charged particles irradiation is of main concern (especially concerning the flux effect and defects generation, for which different mechanisms can be considered) ;
- Predictive numerical simulation have to integrate more accurately the information related to primary damage ;
- Online characterization and in situ deformation under irradiation considering a corrosive environment is essential (solutions already exist but have to be improved sometimes).

Main comments and outcomes of session 2: Primary Damage & Structural defects

Chair: F. GAO, Dept. of Nucl. Eng. and Radiological Sciences, College of Engineering, University of Michigan (Ann Arbor, Michigan, USA)
Co-chair : J.-P. CROCOMBETTE, CEA-DEN-DMN, Service de Recherches de Métallurgie Physique, SRMP (Saclay, France)

Method to model irradiation damage

Different approaches exist to simulate primary damage state:
- Molecular dynamic, as presented in F. Gao's talk, which is good qualitatively and quantitatively to describe the primary state of damage in terms of structure and repartition of defects but quite heavy and quite limited in energy. It works either with empirical potentials with the weaknesses of such

potentials or, more recently with the use of ab initio simulations, addressing charge transfer and charge-modified migration energy of defects. This might be important for ceramics and glass, but ab initio MD limits nowadays to 1000 atoms.
- The Binary Collision Approximation (BCA) approach, as in L. Lunéville talk. This method is very generic and very good at producing number for the amount of damage; but does not describe the defects created.
The work presented by J.-M. Delaye, exemplified an ad hoc method which relies on completely different principles. Indeed this approach does not model at all the irradiation itself but focuses on its effect on the structure of the material. As it appeared during the discussion, such methodology can work only in specific situations and in cases where experimental results are available to valid the defective, port-irradiation structure.

Effect of electronic loses, their importance and the difficulty to model them

Electronic losses appeared clearly in the talks of K. Yasuda on oxides and Y. Ngono-Ravache on polymers. They have shown that such electronic effects are of great importance for their materials. There is no surprise these were experimental talks! But in a sense, electronic losses were also somehow there, as a ghost, in F. Gao's talk, as these electronic losses are known to be of importance in SiC material, but we have little way to model them, beyond the charge transfer effects mentioned at the beginning. There was an agreement that we have a lot of work ahead of us on these matters, both experimentally and theoretically, even if the situation is less dramatic than thought by many. Indeed Monte-Carlo codes for the transport of particles (such as GEANT-4) exist which allow a description of the travelling of excited electrons in matter. Moreover two-temperature models are an emerging tool to describe the transfer of energy from the atomic and electronic systems. Nevertheless no formalism exists to evaluate to creation of defects by excited electrons. The description of swift heavy ion damage or fission track damage in ceramics also involves a lot of works ahead of us. Of course, Molecular Dynamics simulations of heat spikes provide an ad hoc method to simulate heavy ion damage or fission track damage in ceramics. The effectiveness and correctness of this approach remains an open question.

Modeling of neutron with ion

This topic was tackled from two different perspectives by L. Lunéville and J.-P. Mardon. This is a very complex topic we had no time to discuss in details. Questions pertaining to this subject include: in which way, the ion irradiation is similar to neutron irradiation. Do we understand the differences between neutron and ion irradiations? How can we model neutron irradiation and ion irradiation?
J.-P. Mardon was very satisfied with his ion irradiation experiments and he said he has gained understanding on what happens with real neutron irradiations.

A few other points were tackled during the discussion:

The importance of a correct description of primary damage was debated as well as the possibility to effectively measure it. Some thought it is impossible to measure while model laser experiments were suggested by others, as well as experiments with advanced proton sources. The accuracy and utility of the NRT dpa was also discussed in connection with the on-going thinking of the EAN and IAEA on that subject. The fact the dpa was a measure of damage not defects and that the number of defects is roughly one third of what is predicted by NRT was stressed. Finally the importance of measuring and calculating the sputtering due to fission fragments in nuclear fuels was stressed. This sputtering comes mainly from electronic loses of high energy fission fragments and is due to the heating along the fission track.

Main comments and outcomes of session 3: Evolution & Characterization of the Microstructure

Chair: M. J. CATURLA, Dept. Física Aplicada, Facultad de Ciencias, Fase II, University of Alicante (Alicante, Spain)
Co-chair: G. JOMARD, CEA-DEN-DEC, Service d'Etudes et de Simulation du Comportement des Combustibles, SESC (Cadarache, France)

The main topic that has been addressed was related to the link between modelling and experiments. It is clear that both approaches must take benefit from a strong coupling (not only comparison) between experimentalists and modelers. Validation of models is not always straightforward since there are only rare cases where calculations results and experimental observations are directly comparable: Positron lifetime in defected materials and XAS spectra are two of these. Usually experiments and simulations are not performed at the same size or time scales which complicates the comparison.

Often the interpretation of the experiments requires some modeling. This is the case of positron annihilation experiments where accurate DFT calculations of lifetimes are very useful in the interpretation of the results. In the case of TEM, quantification of the results is complex since it depends on the resolution limit, on the families of loops that can or cannot be observed for a given orientation or even the distance of defects to the surface. In this case maybe TEM image simulations could help in the quantification of the results. Some other methods were also pointed out such as XAFS or XDR which would need data from simulations. It was pointed out that both in the US and in Europe there are some developments towards sharing raw data between scientists, although it is not clear how this will be done in practice. Each community however agrees that the interaction between experiments and simulations must be strengthened.

The issue of the quantification of damage induced by irradiation has been addressed. It seems reasonable to treat dpa for what it is: a measure of the energy deposited in a sample during irradiation instead of a measure of an amount of defects. However, there is still a debate about if this value is being used properly in different contexts.

Among quantities that can be obtained from experiments and which will be very valuable for modelers, the threshold energy for U and O displacement in UO_2 has been identified as a key parameter that needs to be measured.

Main comments and outcomes of session 4: Impact on Physical, Chemical, Mechanical Characteristics & Properties

Chair: I. M. ROBERTSON, Dept. of Materials Science and Engineering, University of Wisconsin-Madison, (Madison, Wisconsin, USA)
Co-chair: D. FÉRON, CEA-DEN-DPC, Service de la Corrosion et du Comportement des Matériaux dans leur Environnement, SCCME (Saclay, France)

The first talk in this session by I. Robertson et al., described experimental observations of dislocations interacting with grain boundaries in un-irradiated and ion-irradiated FCC metals. It was demonstrated that the slip system activated to transfer slip across a grain boundary, at least at low level of strains, was determined by the magnitude of the Burgers vector of the dislocations left in the grain boundary as a result of the transmission event. It was claimed that the local resolved shear stress on the selected slip system needs to be sufficient to move dislocations from the grain

boundary source into the grain. This makes the magnitude of the local resolved shear stress important in irradiated alloys as it must be sufficient to move dislocations through the field of irradiation-produced obstacles. An outstanding question to be answered is does the slip transfer process across grain boundaries change as the local strain is increased.
The second talk by M. Sauzay et al., provided state-of-art simulations and modeling of the impact of the slip bands, cleared channels, impacting a grain boundary and used a critical normal stress and fracture energy criteria to deduce the conditions under which the grain boundary would fail. Important advances in the formulation of the problem are the consideration of the width of the channel, which relates to having dislocations available on not one but multiple parallel and closely spaced slip systems. The cohesive strength of the grain boundary was assumed to change in the near surface region due to oxidation. An interesting question to consider is can this method be used to explain why only select slip band intersections of the grain boundary result in cracking of the grain boundary.
The third talk by A. Courcelle et al., demonstrated that a cold-worked Ti-stabilized austenitic stainless steel exhibits a high resistance to void swelling even at high doses. A comparison of the precipitate interactions with vacancy clusters showed differences between neutron and ion irradiated materials. This work demonstrated the need for further understanding the precipitation so that it may be accounted for appropriately in models.
The fourth talk by G. Monnet described the complex problem of building a multi-scale model of radiation-induced hardening was presented. It demonstrated the advances that have been made as well as those needed to advance the field further. For example, a challenge is to identify the dominant processes responsible for a particular effect, e.g., the mechanisms of annihilating the radiation-produced damage as it is known that the annihilation event does not produce an easy glide path, which raises the question of the nature of the remnant damage. Also, as indicated by the talk by M. Sauzay, the cleared channels involve activity on closely spaced parallel planes within a confined volume. An interesting question is how to move from isolated behaviors and interactions, which are well understood, to the more complex and important ones involving the collective behavior or summation of all possible interactions and use this information to bridge between length scales.
The fifth talk by J. Noirot et al., described the impact of high burn up on fuel behavior. Significant changes in the microstructure, grain size reduction, larger bubbles and precipitate, were shown to occur in the high burn-up region. It was suggested that the differences that have been reported may be due to the different levels of impurities in each of the systems.
The last talk by S. Perrin et al., was dedicated to the influence of irradiation on stainless steel corrosion in PWR primary conditions. A focus was made on damages due to IASCC in austenitic stainless steel 316L in order to understand the role of defects on oxidation process. Oxide layer characterizations (oxide layer properties, oxide kinetic, oxygen transport) were performed on non-implanted and implanted samples with xenon. This study shown that, the defects created by xenon implantation have an effect on the oxide layer thickness and composition, on the internal layer growth kinetic and on the oxygen diffusion in this layer.

Key points from the discussion:

1 - Multi-scale modelling and simulation efforts that are coupled directly to a multi-scale experimental efforts are needed. The complexity of the irradiation damage produced and the interaction of dislocations either singly or collectively with the radiation-produced damage presents a major challenge in the development of a multi-scale model that links material performance to atomic- and meso-scale events that can occur over disparate timescales. Despite the advances that have been made there remains a significant gap in the spatial and temporal scales accessible to both approaches. Continued efforts to reduce the gap are needed or, alternatively, the problem being considered must be carefully framed to allow for the synergistic coupling of the two approaches.

On the experimental side, what is needed are in-situ in-operandi experimental methods that bridge the appropriate spatial and temporal scales that will provide new insight into the governing processes and mechanisms. For example, the closed loop system with the in-situ analytical tools that is being construction should provide new insight to the governing mechanisms of corrosion. Coupling this experimental capability with advances in computational methods provides a unique opportunity to address this critical problem.

On the computational side, it is necessary to continue to improve the algorithms to capture the new physics and chemistry that will permit studies of performance of materials in environments that simulate actual operational conditions. Also, needed are advances in the potentials used in simulations such that they more accurately reflect and actual material, e.g., magnetic effects and native oxides.

2 - Date-enabled and –driven science and engineering. The nuclear materials community has over the decades generated a significant amount of data and knowledge, much of which will be lost with the generational change in nuclear material scientists and engineers. There is a unique opportunity to create searchable data bases that will allow new interrogation approaches to be employed to look for hidden synergies and effects. The value of moving in this direction was debated with no consensus being reached. The discussion did raise the issue about gathering and storage of data that would be meaningful to others.

Other challenges were also mentioned:
- From atomic to the continuum scale, modelling and characterization of damages need to be improved on RPV and stainless steels, on new cladding alloys, on fuels… ;
- While modelling and observation scales in terms of geometrical dimensions are quite in accordance, it is not at all the case in terms of time. The accumulation of irradiation or environmental damages is over years while atomistic modelling is often much less than a second. The resolution of this dilemma may lead to the evaluation of incubation and initiation periods of damage phenomena ;
- Irradiation is one of the parameters among other ones coming from the environment. It is not only multiscale phenomena which are studied in geometrical and time scale, but also multi-parametric processes.

DE LA RECHERCHE À L'INDUSTRIE

cea den

instn

MINOS

Centre of Excellence for Nuclear Materials

2nd International Workshop

Irradiation of Nuclear Materials: Flux and Dose Effects

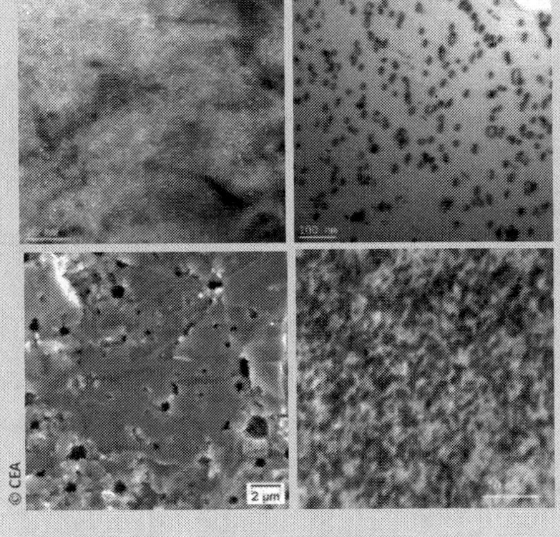

November 4-6, 2015

CEA – INSTN Cadarache, France

MATERIALS

AREVA

Materials Simulation & Engineering

Réacteur Jules Horowitz Reactor

RJH Reactor under construction at the CEA Cadarache

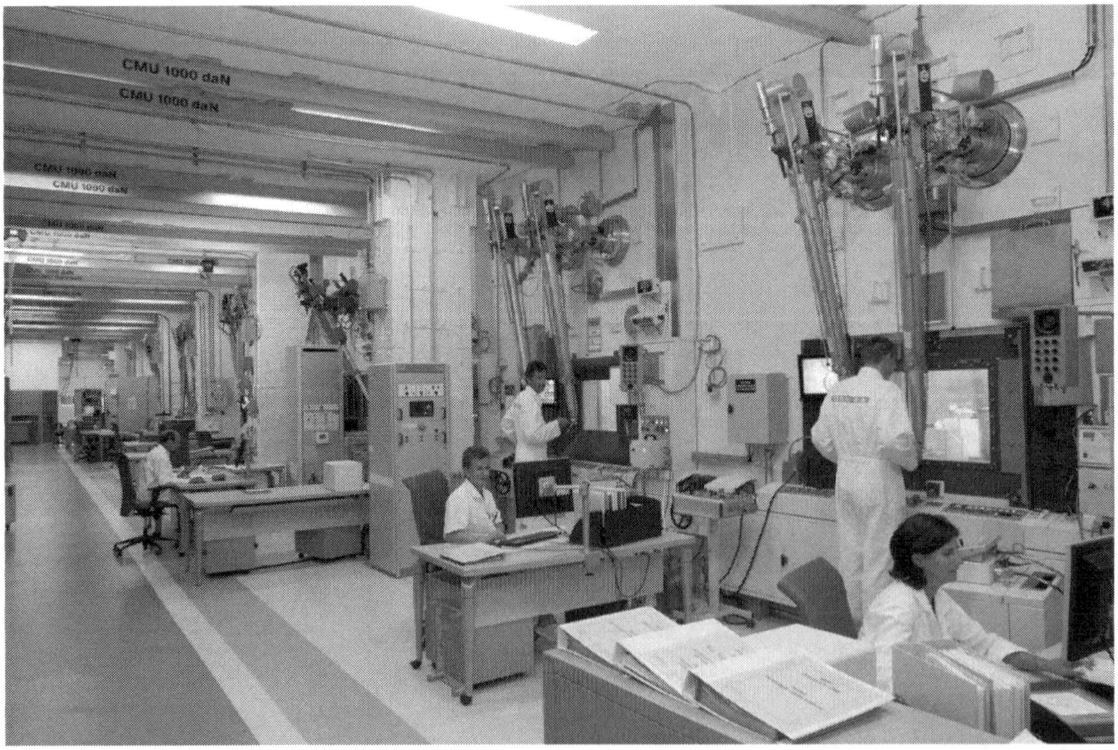

Front area of the LECA hot facility (Fuel Research Department, CEA Cadarache)

MINOS
2nd workshop international
4-6 nov. 2015, (CEA- INSTN, Cadarache)

edp sciences EPJ Web of Conferences

Statement of Peer review

In submitting conference proceedings to EPJ Web of Conferences, I certify to the Publisher that I adhere to the **Policy on Publishing Integrity** of the journal in order to safeguard good scientific practice in publishing.

1. All articles have been subjected to peer review administered by the proceedings editors.
2. Reviews have been conducted by expert referees, who have been requested to provide unbiased and constructive comments aimed, whenever possible, at improving the work.
3. Proceedings editors have taken all reasonable steps to ensure the quality of the materials they publish and their decision to accept or reject a paper for publication has been based only on the merits of the work and the relevance to the journal.

Title, date and place of the conference

MINOS 2nd International Workshop,
Irradiation of Nuclear Materials: Flux and Dose Effects
November 4-6, 2015, CEA - INSTN Cadarache, France

Proceedings editor(s):

Christophe GALLÉ
CEA Saclay
Nuclear energy Division
Department of Nuclear materials
MINOS Centre of Excellence for Nuclear Materials
91191 Gif-sur-Yvette, Cedex France

Date and editor's signature

February 25, 2016

EPJ Web of Conferences **115**, 01001 (2016)
DOI: 10.1051/epjconf/201611501001

MINOS
Centre of Excellence for Nuclear Materials

2nd Int. Workshop Irradiation of Nuclear Materials: Flux and Dose Effects
November 4-6, 2015, CEA – INSTN Cadarache, France

EMIR Federation: Network of Accelerators Dedicated to Material irradiation

Serge BOUFFARD, EMIR Federation and CIMAP

EMIR Federation, CIMAP-GANIL, CEA-CNRS-ENSICAEN-UNICAEN (Caen, France)

The ageing under irradiation of the materials involved in nuclear power industry is a well known problem which has been extensively studied over several decades. These studies driven by industrial requirements cover all the continuum of the approaches from atomic scale to the technologic tests. All along the nuclear fuel cycle, materials are damaged by different particles: chemical additives in fuel synthesis, the components of nuclear reactors, materials for the waste disposals and of course the nuclear fuel. In all cases, irradiation induced aging of the materials is a very important factor to be taken into account for safety reasons and economic considerations. These are the reasons why the study of materials under irradiation remains a major concern.

Understanding the processes of the damage of metals has made significant progress in recent years thanks to the development of the simulation and new experimental approaches. However, much remains to be done to obtain predictive simulations of the industrial materials in operating conditions. Moreover, for iono-covalent ceramics, knowledge on the dynamic of defects remains much patchy. For these materials, an important effort must be done at all temporal and spatial scales to reach the knowledge level of metallic alloys. This must be also done for the organic materials.

Despite this, the French research potential in the field of materials under irradiation remains relatively low. Given this relative weakness of the means of irradiation and online characterizations, six accelerator platforms have created a network with the objective to coordinate their instrumental developments and the access to the experimental facilities. This network which belongs to many entities, has recently obtained the CNRS label of federation: federation of accelerators dedicated to material irradiation. This network consists of 6 platforms with unique equipments and specific expertises. Its objectives are the followings:

- promote experiments on materials under irradiation,
- networking the irradiation facilities,
- organize the service for external users,
- promote the developments of facilities in accordance of research evolution,
- scientific animation and training on material irradiation
- inform the organizations about these researches

The six platforms are:

-	JANNUS	Saclay	CEA/DEN/DMN/SRMP;
-	JANNUS	Orsay	CSNSM CNRS/IN2P3 – Paris-Saclay University;
-	CEMHTI	Orléans	CNRS/INC;
-	SIRIUS	Palaiseau	LSI CEA/DSM/IRAMIS, CNRS/INP, Ecole Polytechnique ;
-	HVTEM	Saclay	CEA/DEN/DMN/SRMA;
-	CIMAP/GANIL	Caen	CEA/DSM/IRAMIS, CNRS/INP, ENSICAEN, Caen University

Their specificities and their online characterization devices will be described as well as the statistics on the experimental proposal (distribution on requested beam, on subjects...)

The table below gives the main specificities of the platforms.

	specificities	In-situ characterizations
Ion accelerators		
JANNuS – Saclay – SRMP		
triple beams	irradiation and implantations simultaneously	in-situ RAMAN and RBS, NRA, ERDA
JANNuS – Orsay – CSNSM		
double beams	irradiation or implantation and TEM observation simultaneously	in-situ TEM and RBS
CEMHTI – Orléans		
high-energy light ions	large range	in-situ RBS and variable temperature Raman, mechanical stress
CIMAP – GANIL – Caen		
swift heavy ions	large range effects dominated by electronic excitations	in-situ X-ray diffraction, infrared and UV-visible spectroscopies, gas emission
Electron accelerators		
SIRIUS – LSI – Palaiseau		
2.5 MeV electron	large irradiated area energy range	in-situ : electric, magnetic, optic properties
HVTEM – SMRA – Saclay		
1 MeV electron	Irradiation inside a high voltage TEM	in-situ imaging of structural evolution, environmental electron microscope

EMIR network organizes an annual call for experimental proposals, the proposals are assessed by two external referees and an international scientific committee ranks the proposal from these assessments and their own opinion.

For more information: http://emir.in2p3.fr/

The EMIR facilities (from left to right): Jannus-Saclay, HVTEM, CEMHTI, CIMAP-GANIL, SIRIUS, Jannus-Orsay

EPJ Web of Conferences **115**, 01002 (2016)
DOI: 10.1051/epjconf/201611501002

MINOS
Centre of Excellence for Nuclear Materials

2nd **Int. Workshop** Irradiation of Nuclear Materials: Flux and Dose Effects
November 4-6, 2015, CEA – INSTN Cadarache, France

Surrey Ion Beam Centre: the EPSRC "MRF" for Ion Beam Applications

Roger P. WEBB[1]

[1]Surrey Ion Beam Centre, University of Surrey, (Guildford, UK)

The Surrey Ion Beam Centre (SIBC) has been sponsored as a National Centre for research in the UK since 1978. It underpins more than £40M of active UK research grants at any one time on a range of subject areas but mostly to support micro-electronics radiation damage studies. It was part of large European infrastructure project (SPIRIT) providing access to ion beam equipment across Europe. In 2002 it was awarded the Queens Anniversary prize for research achievement in Ion Implantation and Optoelectronics. The citation which accompanies the award says "*For over thirty years the University has been recognized as a pre-eminent institution in the field of ion implantation and optoelectronics. Its innovative research into these key technologies has had an outstanding impact on the development of the modern electronics industry worldwide.*" The SIBC works actively with industry, developing bespoke processes and services, particularly for the photonics industry and provides ion beam facilities to ~20 companies across the world. It operates a stringent quality control program and is one of the few ion beam laboratories in the world to operate under ISO 9001 certification.

The SIBC has become one of the world leaders in the field of ion beam research. In the 1970's and early 80s it pioneered the rapid thermal annealing technique[1] which is now universally used by the semiconductor industry, in the 80's it developed [2,3] "radiation hard" silicon on insulator substrates for space and military applications and for a time was the world's only supplier of these substrates. In the 1990s it developed DataFurnace [4] a computer algorithm to enable quantified analysis using ion beam techniques (one of 2 recognised by the IAEA). The Ion Beam Analysis DataFurnace is a computer code to extract *elemental depth profiles* from Rutherford backscattering and related ion beam analysis spectra. It is able to solve the inverse problem ("given the spectrum, what is the profile") *automatically*, without user intervention. In the 1990's the SIBC also began its work on silicon photonics, beginning with research on $\beta FeSi_2$ [5] in silicon and later combing this work with defect engineering [6]. This led to the formation of a spinout company SiLight Technologies and a prestigious European Research Council grant.

More recently in the field of microelectronics the SIBC has worked on novel approaches to activate dopants for shallow junction devices [7]. The SIBC has also looked to widen its approach and apply its techniques to new areas of research. This has included the biomedical applications of ion beams [8], where a network led by the SIBC was instrumental in bringing the clinical community together to make a case for particle therapy in the UK – three new proton therapy units are now in the process of being built in the UK after much lobbying. With colleagues in Japan, the IBC has pioneered the development of a new analytical technique – MeV SIMS [9-10], this has the potential to produce high molecular mass concentration maps in ambient pressure conditions with high spatial resolution. This is not currently possible using any other technique. The IAEA are involved in promoting and developing this technique worldwide. Recently the IBC has used the capabilities afforded by its DataFurnace algorithm to initiate work with Forensic service providers, the police and Home Office and similar establishments across mainland Europe. This has led to ion beam techniques being recognised as potentially of use for high impact crime cases. In 2007 the IBC received ISO9001 accreditation and is now actively pursuing ISO 17025 certification for its Ion Beam Analysis protocol.

Total IBA [11] has now been developed from DataFurnace and is a very powerful new ion beam analysis characterisation technique enabling the unambiguous depth profiling of very complex thin films, and in principle making chemical tomography of small samples feasible. This new method, which we have been instrumental in developing, fully exploits the complementary information available from the photon and particle methods.

Over the years the SIBC has developed its portfolio of industrial users. Work with these companies is done via means of a non-disclosure agreement. What the company wishes to achieve is first established and then a process is developed to meet these needs. Sometimes the process is then shipped out to the company in question but more often, batches of wafers are sent to the SIBC for the bespoke implantations and processing required. A turnaround time of typically 3-5 days is achieved meaning that the IBC supplies expert services to companies across the world. The SIBC also supplies bespoke consultancy when required. It runs with a relatively robust business model attracting approximately equal funding from UK research councils, the EU and industry.

The Vertical Beam Line for Cell Irradiation

Semiconductor end station

The Ion Implantation Facility & Clean Room

Beam lines from 2MV Tandem

References

[1] B.J.Sealy, "Rapid Thermal Annealing of Ion Implanted Semiconductors", *Nuclear Physics Applications on Materials Science*, **144**, NATO ASI Series (Springer, Eds Recknagel, Soares), pp215-238, 1982.

[2] S.S.Gill, I.H.Wilson, "RBS Analysis of Oxide Layers formed by Ion Implantation into Single Crystal Silicon" *Thin Solid Films* **55**, 435-448, 1978.

[3] P.L.F.Hemment, E.A.Maydell-Ondrusz, K.G.Stephens, J.B.Butcher, D.Ioannou, J.Alderman, "Formation of Buried Insulating Layers in Silicon by the Implantation of High Doses of Oxygen", *Nuclear Instruments and Methods*, **209/10**, 157, (1983).

[4] N.P.Barradas, C.Jeynes, R.P.Webb "Simulated annealing analysis of Rutherford backscattering data", *Applied Physics Letters*, **71**, 291 (1997).

[5] D.Leong, M.Harry, K.J.Reeson, K.P.Homewood, "A Silicon/Iron-Disilicide Light-Emitting Diode Operating at a Wavelength of 1.5 μm", *Nature*, 1997, Vol.387, No.6634, Pp.686-688.

[6] Wai Lek Ng, M. A. Lourenco, R. M. Gwilliam, S. Ledain, G. Shao & K. P. Homewood An efficient room-temperature silicon-based light-emitting diode Nature, 2001, Vol 410, 192-194

[7] D.Giubertoni, G.Peppon, M.A.Sahiner, S.P.Kelty, S.Gennaro, M.Bersani, M.Kah, K.J.Kirkby, R.Doherty, M.A.Foad, F.Meirer, C.Streli, J.C.Woicik, P.Pianetta, "Deactivation of submelt laser annealed arsenic ultrashallow junctions in silicon during subsequent thermal treatment", *Journal of Vacuum Science and Technology B*, **28(1)**, C1B1-C1B5, 2010

[8] N.G.Burnet, K.J.Kirkby, N.F.Kirkby, "Biomedical Applications of High Energy Ion Beams", *Clinical Oncology*, **16**, 384-385 (2004).

[9] B.N.Jones, V.Palitsin, R.P.Webb, "Surface Analysis with High Energy ToF-SIMS Measured in Parallel with PIXE and RBS", *Nuclear Instruments and Methods B*, **268(11-12)**, 1714-1717, 2010.

[10] Y.Wakamatsu, H.Yamada, S.Ninomiya, B.N.Jones, T.Seki, T.Aoki, R.P.Webb, J.Matsuo, "Biomolecular emission by swift heavy ion bombardment", *AIP Conference Proceedings*, **1321**, 233-236 (2010).

[11] C.Jeynes, M.J.Bailey, N.J.Bright, M.E.Christopher, G.W.Grime, B.N.Jones, V.V.Palitsin, R.P. Webb, "Total IBA – Where Are We?", *Nuclear Instruments and Methods B*, **269**, 2251-2253, 2011.

EPJ Web of Conferences 115, 01003 (2016)
DOI: 10.1051/epjconf/201611501003
© Owned by the authors, published by EDP Sciences, 2016

MINOS Centre of Excellence for Nuclear Materials 2nd Int. Workshop Irradiation of Nuclear Materials: Flux and Dose Effects
November 4-6, 2015, CEA – INSTN Cadarache, France

The Jules Horowitz Reactor Research Project: A New High Performance Material Testing Reactor Working as an International User Facility – First Developments to Address R&D on Material

Gilles BIGNAN[1], Christian COLIN[1], Jocelyn PIERRE[1], Christophe BLANDIN[1], Christian GONNIER[1], Michel AUCLAIR[2], Franck ROZENBLUM[2]

[1] CEA-DEN-DER, JHR Project (Cadarache, France)
[2] CEA-DEN-DRSN, Service d'Irradiations en Réacteurs et d'Etudes Nucléaires, SIREN (Saclay, France)

The Jules Horowitz Reactor (JHR) is a new Material Testing Reactor (MTR) currently under construction at CEA Cadarache research center in the south of France. It will represent a major research infrastructure for scientific studies dealing with material and fuel behavior under irradiation (and is consequently identified for this purpose within various European road maps and forums; ESFRI, SNETP…). The reactor will also contribute to medical Isotope production.

The reactor will perform R&D programs for the optimization of the present generation of Nuclear Power Plans (NPPs), will support the development of the next generation of NPPs (mainly LWRs) and also will offer irradiation capabilities for future reactor materials and fuels.

JHR is fully optimized for testing material and fuel under irradiation, in normal, incidental and accidental situations:

- with modern irradiation loops producing the operational condition of the different power reactor technologies ;
- with major innovative embarked in-pile instrumentation and out-pile analysis to perform high-quality R&D experiments ;
- with high thermal and fast neutron flux capacity and high dpa rate to address existing and future NPP needs.

JHR is funded and steered and will be operate as an international user-facility open to international collaboration. This lead to the following topics:

- the existence of an international consortium gathering the funding organizations to steer the project ;
- the setting-up of an international scientific community around JHR through seminars, working groups to optimize the experimental capacity versus future R&D needs ;
- the preparation of the first JHR International Program potentially open to non-members of the JHR consortium.

It will answer needs expressed by the scientific community (R&D institutes, TSO…) and the industrial companies (utilities, fuel vendors…). Consequently, the JHR facility will become a major scientific hub for cutting edge research and material investigations (multilateral support to complete cost effective studies avoiding fragmentation of scientific effort, access to developing countries to such state of the art research reactor facilities, supra national approach….).

Considering material behavior under irradiation, such studies is most of the time associated with a complex multi-physical modelling of the materials' behaviors. It requires well controlled and instrumented irradiation experiments in material testing reactors.

This paper gives an up-to-date status of the construction (Fig. 1) and of the developments performed to build the future experimental capacity dedicated to the material irradiations in JHR reactor. In-core and in reflector devices will be presented (Fig. 2), corresponding to large ranges of temperature and neutrons flux for the irradiation conditions. A special attention focuses on the improvement of the thermal stability and gradients of the interest zones in samples despite strong gamma heating and on an improvement of the instrumentation devoted to the experiments. Some specific devices in support of LWR type reactors will be described such as equipment designed for the qualification of Reactor Pressure Vessel (RPV) steels, for the study of the stress corrosion cracking assisted by irradiation phenomena (IASCC), or for the studies of creep-swelling of structural materials.

Fig. 1: General view of the JHR building (September 2015).

Fig. 2: reactor pool (left) and part a reactor block (right) of the JHR (September 2015).

References

[1] G. Bignan, X. Bravo, "The Jules Horowitz Reactor: A new high performance MTR (Material Testing Reactor) working as an International User Facility in support to Nuclear Industry, Public Bodies and Research Institutes", Nuclear Energy International Journal, December 2014, pp. 26-30.

EPJ Web of Conferences **115**, 01004 (2016)
DOI: 10.1051/epjconf/201611501004

MINOS
Centre of Excellence for Nuclear Materials
2nd Int. Workshop Irradiation of Nuclear Materials: Flux and Dose Effects
November 4-6, 2015, CEA – INSTN Cadarache, France

The Characteristics and Irradiation Capabilities of MARIA Research Reactor in NCBJ Świerk

Grzegorz KRZYSZTOSZEK
NCBJ, Department of Nuclear Energy, Świerk, (Poland)

1. MARIA research reactor description

The high flux research reactor MARIA is a water and beryllium moderated reactor of a pool type with graphite reflector and pressurised channels containing concentric six-tube assemblies of fuel elements. It has been designed to provide high degree of flexibility. The fuel channels are situated in a matrix containing beryllium blocks and enclosed by lateral reflector made of graphite blocks in aluminium cans. The MARIA reactor is equipped with vertical channels for irradiation of target materials, a rabbit system for short irradiations and six horizontal neutron beam channels. The MARIA reactor reached its first criticality in December 1974. The reactor was in operation until 1985 when it was shut down for modernization. The modernization encompassed refurbishment and upgrading of technological systems such as:

- enlargement of beryllium matrix,
- inspection of the graphite bocks,
- upgrading of ventilation and temperature systems.

The second step of upgrading the technological system was done from 1996-2002 (during regular maintenance) and it was consisted with: replacement of heat exchangers, replacement of instrumenttation and control system, upgrading of radiation protection system, modernization of fuel element integrity monitoring system. The reactor was fully converted from HEU to LEU fuel in the end of August 2014 [1]. In the end of March MARIA reactor received the new license for reactor operation till 2025 year.

The main characteristics and data of MARIA reactor are as follows [2]:

▪ nominal power	30 MW(th),
▪ moderator	H_2O, beryllium,
▪ cooling system	channel type,
▪ fuel assemblies:	
- material	U_3Si_2
- enrichment	19,75%
- cladding	aluminium
- shape	five concentric tubes
- active length	1000 mm.
▪ output thermal neutron flux at horizontal channels	3 - 5 x 10^9 n/cm^{2}s

The main areas of reactor application are as follows: production of radioisotopes, testing of fuel and structural materials for nuclear power engineering, neutron radiography, neutron activation analysis, neutron transmutation doping, research in neutron and condensed matter physics. For today the NCBJ has a program for MARIA research reactor operation till 2030.

2. Characteristics of MARIA reactor for installation loops and rigs

MARIA reactor core contains the fuel assemblies which are installed in pressurized channels embedded in matrix sockets. The matrix is composed of beryllium blocks which are fastened to the support slab in reactor pool on the level +2.75 m [3]. Beryllium blocks of the core matrix as well as the graphite blocks creating the radial reflector are positioned in the sockets of separator slab on the level – 1.4 m (6x8) Fig.1 and Fig. 2.

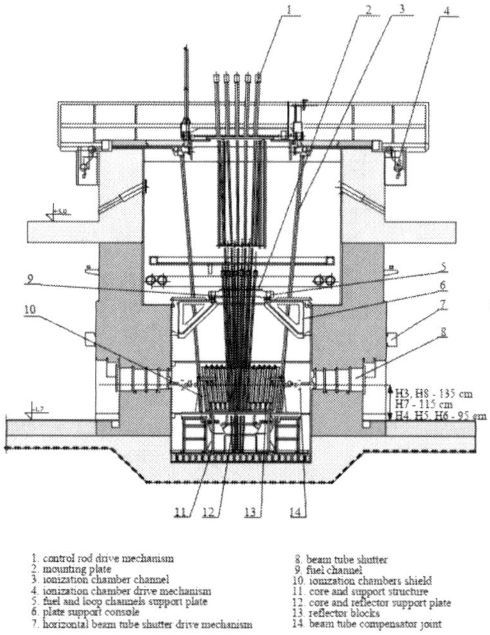

a)

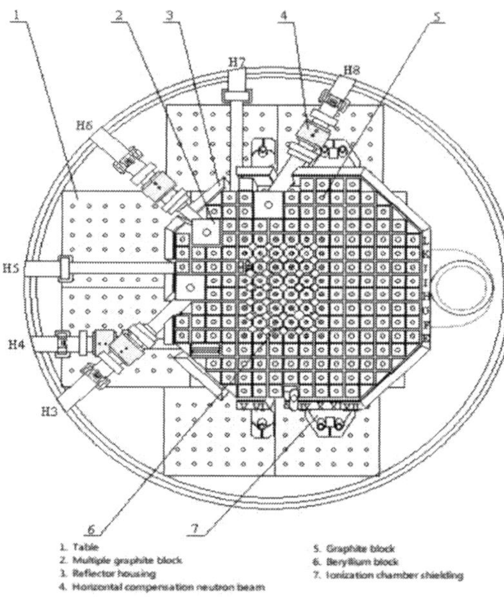

b)

Fig. 1: a) vertical cross-section of the reactor pool, b) top view of the reactor core and reflector.

The possibility of placing of the experimental channel, i.e. channels, rigs and loops depends on the two factors: of the position in the core and arrangement and dimensions of holes in the support slab. The place within in the core can be matched to any experimental device by means of constructing suitable beryllium and graphite blocks.

The neutron-thermal characteristics for the research channels are as follows:

- fast neutron flux $\Phi_f = 1.0 - 1.5 \cdot 10^{14}$ n cm $^{-2}$ s^{-1}
- thermal neutron flux $\Phi_{th} = 1.0 \cdot 10^{14}$ n cm $^{-2}$ s^{-1}
- heat generation $q_{Fe} = 3 - 4$ W $\cdot$ g^{-1}

3. Neutron irradiation services

Neutron irradiation services provided at the MARIA research reactor mainly include radioisotope production. Irradiation services are performed in various facilities in the MARIA reactor, with proper account for the required neutron flux 1, irradiation time, and target mass and geometry. The standard vertical in-core isotope channels as well as special ones equipped with hydraulic transport system are in operation. From 2010 in MARIA reactor started irradiation of uranium plates for Mo-99 production. Irradiations are carried out in two molybdenum channels and irradiated plates are sent to Petten facility in the Netherlands for processing. From September 2014 MARIA offers special irradiation in converter of 14 MeV neutrons.

References

[1] G. Krzysztoszek, Full core conversion from HEU to LEU fuel in MARIA reactor, 9th IAEA TM, Samarkand, Uzbekistan 3-5 June 2015.

[2] K. Pytel et all, Operation Safety Report of MARIA reactor, ERBM-2015.

[3] G. Krzysztoszek, Maria research reactor infrastructure for ICERR, IAEA TM, Aix-en-Provence, 10-12 April 2012.

EPJ Web of Conferences **115**, 01005 (2016)
DOI: 10.1051/epjconf/201611501005

MINOS
Centre of Excellence for Nuclear Materials

2nd **Int. Workshop** Irradiation of Nuclear Materials: Flux and Dose Effect
November 4-6, 2015, CEA – INSTN Cadarache, France

Plasma Material Interaction in Tokamak: the Contribution of WEST and of Laboratory Studies

Christian GRISOLIA[1] and the Tore Supra Team

[1] CEA-DSM-IRFM, Service Intégration Plasma Paroi, SIPP (Cadarache, France)

In 2016, the WEST tokamak will be in operation and the first plasmas realized. This setup is a machine with a full W actively cooled plasma facing components (PFCs). Depending on the heat loads that must be sustained by the WEST PFCs [1], some of them will be composed of massive tungsten welded on heat sink and other of W coatings on carbon tiles (see Fig. 1).

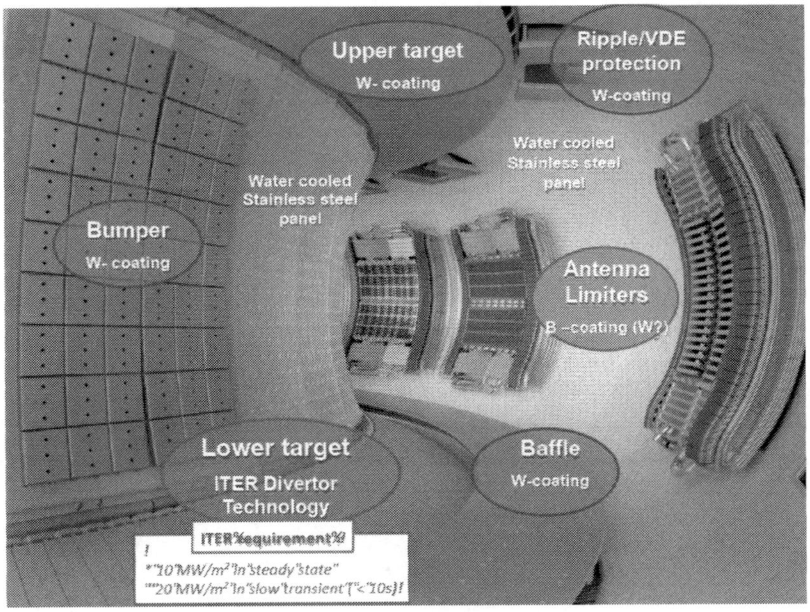

Fig. 1: Summary of the WEST Plasma Facing Component in its final configuration.

During plasma operation, these W PFCs are submitted to harsh heat loads and plasma out flow [2]. The PFCs placed at the lower target of the machine (see figure 1) are ITER type actively cooled components and will be submitted in WEST to heat loads from 10 to 20MW and to particles flux densities up to 10^{23} particles m^{-2} s^{-1}. The ITER technology used here will be described in the presentation as well as the one using coated materials. The comparison of the expected heat and particles loads expected in WEST and in ITER will be also compared.

Several main ITER issues can be studied in such a setup. How the materials evolve when submitted at so high flux? We will study this point and this topic will be described in the presentation.

The WEST can also be used to address one important open issue for ITER that is the deuterium/tritium trapping in massive and dust materials.

A French consortium has been put in place to study at the laboratory scale deuterium and trapping in relevant massive and dusty material. The work-program of such project will be presented during the meeting with the first important results obtained in experimental results [3, 4]. The modelling approach will be also described with some of the key results obtained on tungsten materials [5, 6, 7]. From existing and published results, the consequences of neutrons interaction on the tritium trapping in the ITER machine will be finally discussed.

References

[1] The WEST project: Testing ITER divertor high heat flux component technology in a steady state tokamak environment, J. Bucalossi et a., Fusion Engineering and Design 10/2014; 89(7-8):907-912. DOI:10.1016/j.fusengdes.2014.01.062.

[2] WEST Physics Basis, C. Bourdelle et al., Nuclear Fusion 06/2015; 55(6):063017. DOI:10.1088/0029-5515/55/6/063017.

[3] Dynamic fuel retention in tokamak wall materials: An in situ laboratory study of deuterium release from polycrystalline tungsten at room temperature, R. Bisson, S. Markelj, O. Mourey, F. Ghiorghiu, K. Achkasov, J.-M. Layet, P. Roubin, G. Cartry, C. Grisolia, T. Angot J. Nucl. Mater. (2015), http://dx.doi.org/10.1016/j.jnucmat.2015.07.028.

[4] Tritium absorption and desorption in ITER relevant materials: comparative study of tungsten dust and massive samples, C. Grisolia, E. Hodille, J. Chene, S. Garcia-Argote, G. Pieters, A. El-Kharbachi, L. Marchetti, F. Martin, F. Miserque, D. Vrel, M. Redolfi, V. Malard, G. Dinescu, T. Acsente, F. Gensdarmes, S. Peillon, B. Pegourié, B. Rousseau, Journal of Nuclear Materials, Volume 463, August 2015, pp. 885-888.

[5] Hydrogen diffusion and vacancies formation in tungsten: Densirty Functional Theory calculations and statistical models, N. Fernandez, y. Ferro and D. Kato, acta Materiala 94 (2015) 307-318.

[6] Macroscopic rate equation modeling of trapping/detrapping of hydrogen isotopes in tungsten materials, E.A. Hodille, X. Bonnin, R. Bisson, T. Angot, C.S. Becquart, J.M. Layet, C. Grisolia, accepted for publication, Nucl. Mater. (2015), http://dx.doi.org/10.1016/j.jnucmat.2015.06.041.

[7] Study of Hydrogen Isotopes behavior in tungsten by a multi trapping macroscopic rate equation model, E. A. Hodille, Y. Ferro, N. Fernandez, C. S. Becquart, T. Angot, J. M. Layet, R. Bisson, C. Grisolia, accepted for publication in Physica Scripta, 2015.

EPJ Web of Conferences **115**, 02001 (2016)
DOI: 10.1051/epjconf/201611502001

MINOS
Centre of Excellence for Nuclear Materials
2nd **Int. Workshop** Irradiation of Nuclear Materials: Flux and Dose Effects
November 4-6, 2015, CEA – INSTN Cadarache, France

Modeling and Simulation of Primary Damage and Structure Evolution in Ceramics and Metals

Fei GAO
Dept. of Nucl. Eng. and Radiological Sciences, College of Engineering, University of Michigan (Ann Arbor, Michigan,USA)

Understanding the basics of ion-solid interaction and irradiation damage has led to significant developments in state-of-the-art atomic-level, kinetic Monte Carlo and meso-scale simulations, and these simulations have dramatically advanced the understanding of defect and defect processes in a number of materials, ranging from metals to semiconductors to insulators. Recent progress in multi-scale simulations of fundamental ion-solid interactions, defect generation and defect evolution in ceramics and metals is reviewed, which will provide significant insights into the radiation resistance behavior of these materials.

Recently, large-scale *ab initio* MD simulation methods have been developed for the study of radiation damage in ceramics [1, 2], and these methods are used to explore the effects of charge transfer and charge-density redistribution on the dynamics and charge-state of defect formation. We will focus on the applications of silicon carbide (SiC) because of its a number of understanding physical and chemical properties that lead to its potential applications to semiconductor devices, nuclear reactors for power generation and petrochemical industries, particularly under severe environments. The small cross section, low activation and good thermal conductivity of SiC under neutron irradiation lead to its potential use in structural components for fusion reactor or nuclear waste technology, and as cladding material for gas-cooled fission reactors. Fig. 1(a) shows a perfect structure of 3C-SiC, where large spheres represent Si atoms and small spheres indicate C atoms. Figs. 1(b) and (c) show the charge density of the final configuration of a C recoil (b) or a Si recoil (c) along the {111} direction in the (110) plane. In Fig. 1(b), the most interesting observation is that some charge accumulates on the C vacancy, mainly transferred from the C PKA. The charge is mostly distributed in the region between the C vacancy and one of its neighboring Si atoms, suggesting a tendency for the C vacancy pairing with the Si atom. Obviously, this C vacancy is a positively-charged vacancy. For the Si recoil, the charge density shows that the Si tetrahedral interstitial forms new bonds with its nearest C atoms, as shown by the charge accumulation in the bonding regions (Fig. 1(c)). However, the Si_{TC} prefers to form a new bond with the Si atom below (indicated by an arrow). A strong bond formed between the Si_{TC} and its nearest C atom may facilitate the charge transfer between the two nearby Si atoms, leading to charge accumulation in the bonded regions and the formation of a new covalent bond. Fig. 1(c) also clearly shows that the Si vacancy does not pair with its neighboring atoms, and is in neutral state.

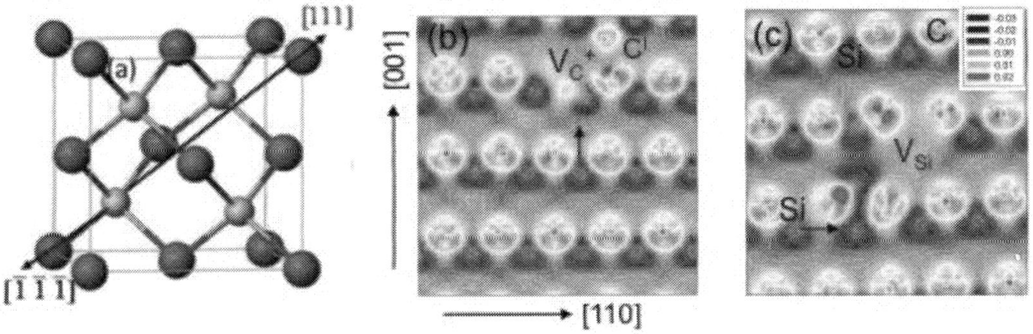

Fig. 1: Perfect crystal of (a) SiC (small spheres-C atoms and large spheres-Si atoms), charge-density difference contour plot for a C recoil (a) and a Si recoil (b) along the $<\bar{1}\bar{1}\bar{1}>$ direction in a $(1\bar{1}0)$ plane (in unit of $|e|/a_b^3$, where ab is Bohr radius).

Next atomic-level study of radiation damage is molecular dynamics simulation that has been approved to be sufficiently powerful that many phenomena associated with radiation damage effects in metals and ceramics. Recent studies of several such phenomena in metals and SiC are reviewed here. In metals, at high enough recoil energy, they create cascades of atomic displacements that result in single and clustered self-interstitial and vacancy defects. These studies provide data on the number of defects produced and their distribution in clusters. I will focus on the formation of defect clusters and dislocation loops. Furthermore, molecular dynamics (MD) simulations to determine defect configurations, defect production, cascade-overlap effects, and defect migration energies and pathways will be reviewed in both bulk SiC and plycrystaslline SiC. The role of interfacial processes in ceramic-ceramic composites will also be discussed. Factors that can improve the radiation tolerance of ceramics will be highlighted. Overlapping of Si cascades results in the interaction of defects and clusters that stimulates cluster growth and produces long-range structural disorder. High-resolution TEM images, disordering behavior, volume change and mechanical properties obtained experimentally and from molecular dynamics simulations are in good agreement, thus providing atomic-level interpretation of experimentally observed features [3].

In terms of polycrystalline SiC, it is of interest to note that the mechanisms of defect generation in nanocrystalline (NC) SiC are found to be different from those observed in metals. However, for average grain diameters larger than 20 nm, most defects are produced within nano grains, rather than inside the grain boundaries in NC SiC. The defect production within grains generally increases with increasing grain size, which is manifested in switching from grain boundary damage to grain damage. The most common defects created in NC SiC are antisite defects, following by vacancies and interstitials, in contrast to those produced in a single crystalline SiC, where the dominate defects are Frenkel pairs. These simulations provide important insights into how nanostructures influence the radiation response or tolerance of ceramics.

A kinetic lattice Monte Carlo (KLMC) model has been developed to investigate the long-time scale simulations associated the recovery and clustering of defects during annealing of a single 10 keV cascade in cubic silicon carbide [4]. The 10 keV Si cascade is produced by MD, and a method of transferring the defects created by MD simulations to the KLMC model is developed. The KLMC model parameters are obtained from MD simulations and *ab initio* calculations of defect migration, recombination, and annihilation. The defects are annealed isothermally from 100 K to 1000 K in the KLMC model. Two distinct recovery stages for close Frenkel pairs are observed at about 200 and 550 K, and the growth of complex clusters is observed above 400 K. These simulation results are in good agreement with available experimental results.

References

[1] F. Gao, H. Y. Xiao, X. T. Zu, M. Posselt, W. J. Weber, "Defect-Enhanced Charge Transfer by Ion-Solid Interactions in SiC using Large-Scale *Ab Initio* Molecular Dynamics Simulations", *PRL 103 (2009) 027405.*

[2] F. Gao, H. Y. Xiao, W. J. Weber, "Ab Initio Molecular Dynamics Simulations of Low Energy Recoil Events in Ceramics", Nuclear Instruments and Methods in Physics Research. Section B, Beam Interactions with Materials and Atoms 269 (2011) 1693.

[3] F. Gao, WJ Weber, "Mechanical Properties and Elastic Constants due to Damage Accumulation and Amorphization in SiC" Phys. Rev. B 69 (2004) 224108.

[4] Z. Rong, F. Gao, and W. J. Weber, "Monte Carlo Simulations of Defect Recovery within a 10 keV Collision Cascade in 3C-SiC", Journal of Applied Physics 102 (2007) 103508.

EPJ Web of Conferences **115**, 02002 (2016)
DOI: 10.1051/epjconf/201611502002

MINOS
Centre of Excellence for Nuclear Materials

2nd Int. Workshop Irradiation of Nuclear Materials: Flux and Dose Effect
November 4-6, 2015, CEA - INSTN Cadarache, France

DART, a BCA Code to Assess and Compare Primary Irradiation Damage in Nuclear Materials Submitted to Neutron and Ion Flux

Laurence LUNEVILLE[1,3], David SIMEONE[2,3]

[1]CEA-DEN-DM2S, Service d'Etudes des Réacteurs et de Mathématiques Appliquées, SERMA (Saclay, France)
[2]CEA-DEN-DMN, Service de Recherches Métallurgiques Appliquées, SRMA (Saclay, France)
[3]Laboratoire de Recherche Conventionné CARMEN (Paris & Saclay, France)

When a material is subjected to a flux of high-energy particles, its constituent atoms can be knocked from their equilibrium positions with a wide range of energies, depending on the exact nature of the collision. The spectrum of damage energy, derived from the exact knowledge of the recoil spectra for each nuclear reaction occurring in the solid, constitutes a vital data set required for understanding how materials evolve under irradiation. The knowledge of such damage energy is relevant to compare the impact of different facilities on the structural behavior and relevant properties of materials.

The DART code [1] was developed for two distinct reasons: the first one was a correct determination of the Primary Knocked on Atoms (PKA) spectrum from reliable cross section data libraries and the second was a crude estimation of the damage energy induced by different irradiations. This last term can be a quick estimation of radiation damage produced in the same material by different nuclear plants and particle accelerators.

Based on the Binary Collision Approximation, this code allows computing the primary spectra produced by neutrons, ions and electrons as well as the damage energy deposited by these particles in a poly atomic material. It is then a tool to compare radiation damage induced in nuclear reactors as well as ion beam facilities.

Figure 1 displays the neutron spectra for different reactors: ITER (fusion reactor), HFIR (High Flux Isotope Reactor), IFMIF (International Fusion Materials Irradiation Facility) and PWR (Power Water Reactor).Two ion beam spectra (Ar 800 keV and Xe 100 keV) are also plotted on this graph.

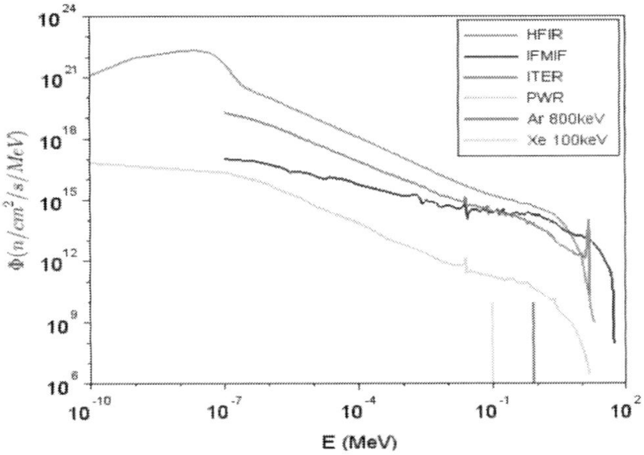

Fig. 1: Neutron spectra for different reactors and beam spectra.

DART calculations were performed to compare damage generated by these irradiations in pure iron. PKA spectra for the different irradiations are plotted on figure 2, showing the strong dependence of the incident spectrum on the energy distribution of recoils.

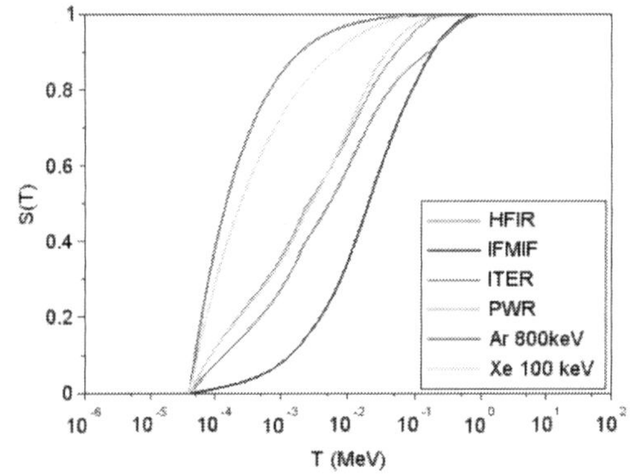

Fig. 2: Comparison of PKA spectra in pure iron for different irradiations using the same formalism for ion and neutron.

As DART allows defining a poly atomic material as a mixture of different isotopes (and not only of natural elements), we achieved the same calculations when adding to pure iron, 10% of a specific isotope of nickel, Ni-59. Figure 3 shows the increase of damage energy due to the presence of this element. Ni-59 increases the damage energy by more than an order of magnitude in HFIR reactor. This effect is related to Ni-59 reaction with thermal neutrons [2]. However, in fusion reactors, the low amount of thermal neutrons leads to a minimal impact on the damage energy due to the presence of Ni-59.

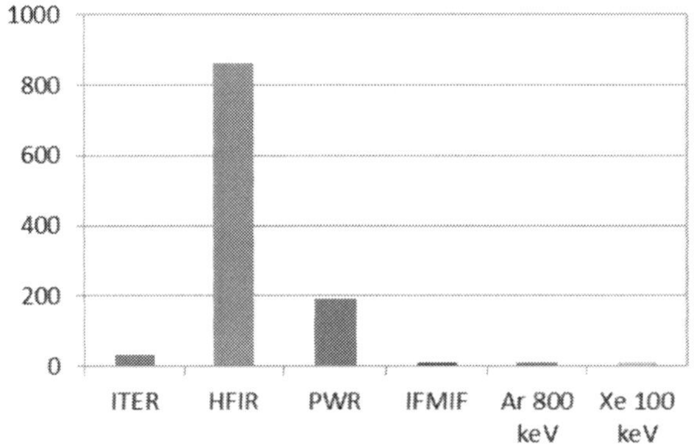

Fig. 3: Increase of the damage energy (in %) when adding 10% Ni-59 in iron (effect of Ni-59 reaction with thermal neutrons).

It can also be pointed out that this result is strongly dependent on the Ni-59 neutron cross sections, showing the crucial role of the cross section data library.

References

[1] L. Luneville, D. Simeone, C. Jouanne, "Calculation of radiation damage induced by neutrons in compound materials", JNM 353, 89-100 (2006).

[2] F. A. Garner, L. R. Greenwood, E. R. Gilbert, M. Griffiths, 2009, "Impact of Ni-59 (n, α) and (n, p) Reactions on Dpa Rate, Heating Rate, Gas Generation and Stress Relaxation in LMR, LWR and CANDU reactors", Proc. 14th Intern. Conf. on Environmental Degradation of Materials in Nuclear Power Systems Water Reactors, Virginia Beach, Virginia, USA, August 23-27, 1344-1354 (2009).

EPJ Web of Conferences **115**, 02003 (2016)
DOI: 10.1051/epjconf/201611502003

MINOS Centre of Excellence for Nuclear Materials — 2nd Int. Workshop Irradiation of Nuclear Materials: Flux and Dose Effects
November 4-6, 2015, CEA – INSTN Cadarache, France

Polymers under Ionizing Radiations: the Specificity of Swift Heavy Ions

Yvette NGONO-RAVACHE[1], Muriel FERRY[2], Stéphane ESNOUF[2], Emmanuel BALANZAT[1]

[1]CIMAP (CEA/CNRS/ENSICaen/UNICaen), CIMAP site GANIL (Caen, France)
[2]CEA-DEN-DPC, Service d'Etude du Comportement des Radionucléides, SECR (Saclay, France)

Compared to materials such as steel or ceramics, organic synthetic polymers are very prone to radiolysis. When submitted to ionizing radiations, molecules composing them are ionized and excited. These initial species lead to the formation of radicals which subsequent recombinations or reorganizations give stable defects. Defects can be either macromolecular or in the form of small molecules evolving from the polymer as gas.

The defects formed under irradiation, and their concentrations, are function of 1) the chemical structure of the polymer repeating unit and 2) the irradiation conditions (temperature, atmosphere, dose, dose rate, and radiation type). Polymers can be submitted to various types of radiations: α, β, γ and ion beams. Detailed studies on the effect of β and γ on polymer ageing do exist, even if some aspects of this should be refined. Conversely, detailed studies on the effect of ion beams, especially Swift Heavy Ions (SHI), are less numerous. Besides, in particular conditions, SHI can be used to simulate the effect of α irradiations on polymers.

Unlike β and γ irradiations that lead to a quite homogeneous energy deposition feature, SHI induce a heterogeneous dose deposition at the nanoscale level. Due to their high LET and because SHI deposit their energy close to the ion path, in a track core of a few nanometers, the local dose nearby these track cores is huge; in between, the dose is very low [1], [Fig. 1].

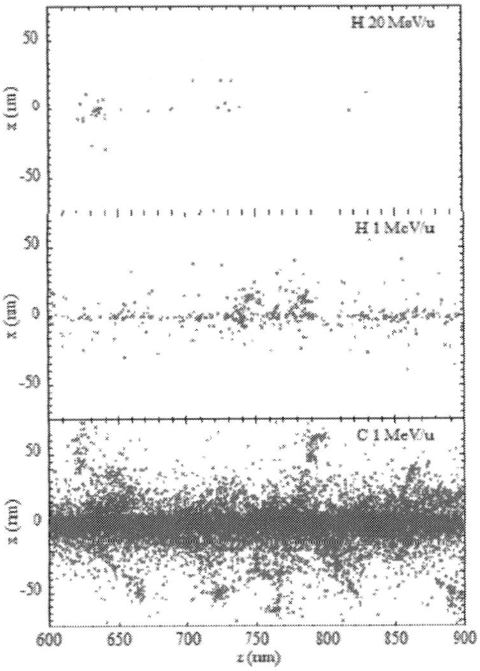

Fig. 1: Energy deposition structure of accelerated ion beams: increase in the density of events, along the track, with increasing LET.

A great number of detailed studies were performed under inert environment to assess the influence of the high ionization and excitation densities induced by SHI on polymer ageing. It was shown that, the huge amount of energy deposited locally by SHI induces specific damage processes [2], which involve complex molecular rearrangements and collective atom motions. Contrary to what has been done under inert environment for assessing the specificity of SHI on polymers, only few detailed studies have been undertaken under oxidative environments. Thus, our studies were aimed to understand how high ionization/excitation densities induced by SHI impact the mechanisms underlying polymer degradation in presence of oxygen.

Polymers submitted to ionising radiations are modified by the creation of macromolecular defects such as unsaturated bonds, chain scissions and crosslinks, or oxidative defects in presence of oxygen. The counterpart of these macromolecular defects is gas emission. We have examined both processes. We have studied the influence of the LET, the dose and the dose rate on different polymers [3]. Additionally, we have considered the influence of the polymer chemical structure on its oxidative ageing, particularly as a function of the LET.

References

[1] Bouffard S, Gervais B, and Leroy C.Basic phenomena induced by swift heavy ions in polymers *NIM B* 105 (1995) 1-4.

[2] Ngono-Ravache Y. Spectroscopic study of the chemical modification induced by swift heavy ions on polymers: the contribution of the CIRIL platform and the CIMAP laboratory. Journal of Physics: Conference Series. Vol. 629 (2015), Art. number: 012006

[3] Ngono-Ravache Y. *et al.* Effect of swift heavy ions on an EPDM elastomer in the presence of oxygen: LET effect on the radiation-induced chemical ageing. Polymer Degradation and Stability, *Volume 111, January 2015, Pages 89-101*

EPJ Web of Conferences 115, 02004 (2016)
DOI: 10.1051/epjconf/201611502004

MINOS Centre of Excellence for Nuclear Materials
2nd Int. Workshop Irradiation of Nuclear Materials: Flux and Dose Effects
November 4-6, 2015, CEA – INSTN Cadarache, France

Structure of Defects and Microstructure Evolution in Oxide Ceramics - Role of Electronic Excitation and Selective Displacement Damage

Kazuhiro YASUDA[1]

[1]Department of Applied Quantum Physics and Nuclear Engineering, Kyushu University (Fukuoka, JAPAN)

Fluorite- and spinel-type oxide ceramics, such as yttria-stabilized zirconia, ceria, magnesium aliminate spinel, are known to be exceptionally resistant to irradiation damage with energetic particles. Those oxide compounds, thereofore, have potential applications to advanced nuclear fuels and innert matrices of transmutaion targets. Fundamental understanding of radiation damage is apparently essential to assess the microstructure stability of fuel/target materials under the hostile environment. In this presentation, structure and stability of radiation-induced defects in the oxide ceramics is reported. Transmission electron microscopy (TEM) and scanning transmission electron microscopy (STEM) is utilized to examine the following topics focucing on role of electronic excitation and selective displacement damage: (1) effects of selective displacement damage of oxygen sublattice in fluorite-type oxides, such as stabilided cubic ZrO_2, CeO_2 [1-3], (2) atomic structure of ion tracks in CeO_2 and $MgAl_2O_4$ induced by swift heavy ions, and microstructure evolution at high fluences under overlapping irradiation of high density electronic excitation damage [4-6], and (3) production and stability of defects in $MgAl_2O_4$ and alumina (α-Al_2O_3) under synergistic irradiation with dispacement damage and electronic excitation [7].

It is noted in fluorite-type oxides that there exists a significant difference in mass between cations and anions, and that the displacement energy of cation-sublattice is larger than the anion one. This leads to a higher displacement damage rate or selective displacement damage in oxygen sublattice in fluorite-type oxides. Figure 1 shows examples of TEM images in CeO_2 showing dislocation loops formed by electron irradiation with energies ranging from 200 to 1250 keV [1]. From contrast analysis and atomic resolution observations with TEM and STEM, those defects are found lying on {111} planes and have been discussed to be composed of solely oxygen ions formed under the selective displacement damage [1]. Similar defects were formed in yttria stabilized ZrO_2, in which entirely different growth process were observed compared to the normal (perfect) dislocation loops [2]. Molecular dynamic simulations in CeO_2, which showed the recovery process of multiple oxygen Frenkel pairs, revealed a tendency of oxygen interstitials for clustering on {111} planes, supporting the interpretation from the TEM and STEM observations [3].

Atomic resolution TEM and STEM observations in $MgAl_2O_4$ and CeO_2 irradiated with 200 MeV Xe ions have shown that the crystal structures are retained at the core region of ion tracks, showing their excellent resistance to high density electronic excitation [4,5]. Figure 2 is an example of atomic resolution image of ion tracks in CeO_2 with annular bright-field (ABF) STEM technique, showing that the crystal structure of Ce-cation column is retained at the core region of ion tracks, whereas O-anion column is preferentially distorted at the core region of ion tracks [5]. The atomic density inside the core damage region of ion tracks in both crystals was found to be decreased, which suggests the existence of high density of vacancies and/or small vacancy clusters inside the ion tracks [5]. Overlap of ion tracks at high fluence was found to develop dislocation structure in both $MgAl_2O_4$ and CeO_2. The development of dislocation structure also results in the formation of small sub-grains in CeO_2. Those microstructure evolution was discussed with the generation of interstitial atoms at the peripheral region of the core damage region of ion tracks [4,6].

It has been shown in $MgAl_2O_4$ and α-Al_2O_3 that synergistic irradiation with displacement damage and electronic excitation retards the nucleation of dislocation loops and enhance their growth [7]. Electronic excitation with 200 keV electrons makes small dislocation loops of interstitial-type unstable in $MgAl_2O_4$ and α-Al_2O_4. Dislocation loops were found to be disappeared during electron irradiation. Analysis of the temperature dependence of the elimination process leads to a conclusion that the dislocation loops are disappeared by dissociating into isolated interstitials. The elimination rate of loops was higher in $MgAl_2O_4$ than α-Al_2O_3, which indicates the high sensitivity of $MgAl_2O_4$ to electronic excitation, and this is considered to play a role for the radiation resistance of $MgAl_2O_4$.

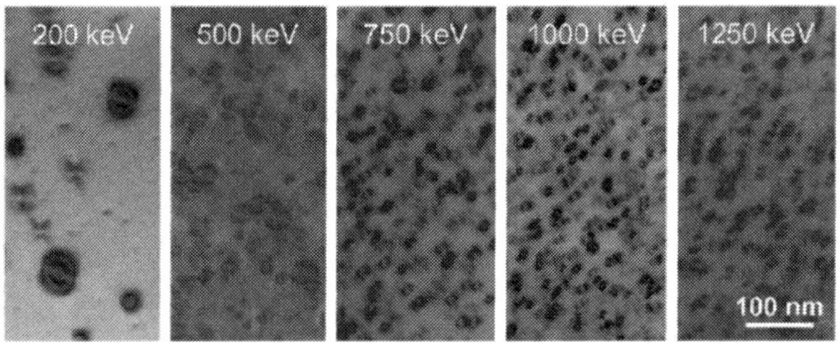

Fig. 1: Dislocation loops formed by electron irradiation ranging from 200 to 1250 keV at room temperature. Dislocation loops are considered to be interstitial-type loops consist of oxygen ions on {111} planes.

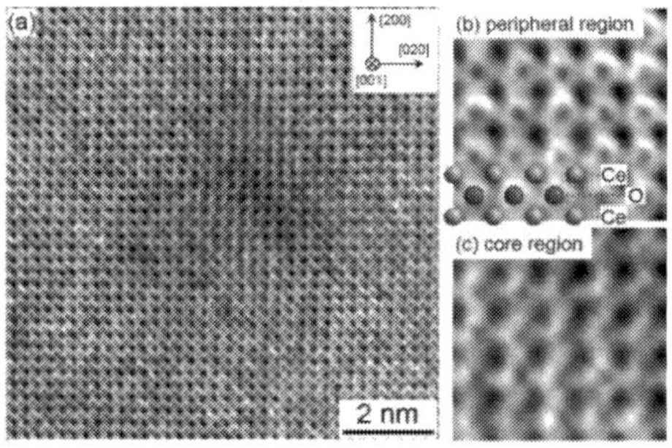

Fig. 2: Annular bright-field (ABF) images of ion tracks in CeO_2 irradiated with 200 MeV Xe ions to a fluence of $3x10^{12}$ cm^{-2}. An ion track is nearly at end-on condition parallel to [001] direction.

References

[1] K. Yasunaga, K. Yasuda, S. Matsumura, T. Sonoda, 'Electron-energy dependent formation of dislocation loops in CeO_2', Nucl. Instr. and Meth. B 266 (2008) 2877.

[2] K. Yasuda, C. Kinoshita, S. Matsumura, A.I. Ryazanov, 'Radiation-induced defect clusters in fully stabilized zirconia irradiated with ions and/or electrons', J. Nucl. Mater., 319 (2003) 74.

[3] K. Shiiyama, T. Yamamoto, T. Takahashi, A. Guglielmetti, A. Chartier, K. Yasuda, S. Matsumura, K. Yasunaga, C. Meis, 'Molecular dynamics simulations of oxygen Frenkel pairs in cerium dioxide', Nucl. Instr. and Meth. B 268 (2010) 2980.

[4] K. Yasuda, T. Yamamoto, M. Etoh, S. Kawasoe, S. Matsumura, N. Ishikawa, 'Accumulation of radiation damage and disordering in $MgAl_2O_4$ under swift heavy ion irradiation', Int. J. Mater. Res., 102 (2011) 9.

[5] S. Takaki, K. Yasuda, T. Yamamoto, S. Mastumura, N. Ishikawa, 'Atomic Structure of Ion Tracks In Ceria', Nucl. Instr. and Meth. B 326(2014) 140.

[6] K. Yasuda, M. Etoh, K. Sawada, T. Yamamoto, K. Yasunaga, S. Matsumura and N. Ishikawa, 'Defect formation and accumulation in CeO_2 irradiated with swift heavy ions', Nucl. Instr. and Meth. B 314 (2013) 185.

[7] K. Yasuda, T. Yamamoto, S. Seki, K. Shiiyama and S. Matsumura, 'Production and Stability of Radiation-Induced Defects in $MgAl_2O_4$ under Electronic Excitation', Nucl. Instr. and Meth. B 266 (2008) 2834.

EPJ Web of Conferences **115**, 02005 (2016)
DOI: 10.1051/epjconf/201611502005

MINOS
Centre of Excellence for Nuclear Materials

2nd Int. Workshop Irradiation of Nuclear Materials: Flux and Dose Effect
November 4-6, 2015, CEA - INSTN Cadarache, Franc

Irradiation Induced Phenomena in Nuclear Glass

Jean-Marc DELAYE[1], Dimitrios KILYMIS[2], Le-Hai KIEU[1], Sylvain PEUGET[1]

[1]CEA-DEN-DTCD, Service d'Etudes et Comportement des Matériaux de Conditionnement, SECM (Marcoule, France)
[2]Laboratoire Charles Coulomb UMR 5221, Université de Montpellier-CNRS (Montpellier, France)

Introduction

In France, non-recoverable radioactive wastes coming from spent nuclear fuels are stored in alumino borosilicate glassy matrices. It is envisaged to store them in a deep geological repository to isolate them from the rest of the environment. These glasses will be subjected to internal irradiations, and after some hundreds of years, to water corrosion due to the arrival of the groundwater in contact with them. Studies are carried out to quantify the radiation and water corrosion effects to characterize the long-term waste package behavior. Experimental and numerical approaches have been considered in order to study the variations of the structural and mechanical properties under irradiation.

Experimentally, it has been shown, on industrial and simplified glasses doped with short lived actinides or irradiated externally by heavy ions, that the elastic effects lead to different structural (decrease of the polymerization glass networks, increase of the disorder) and mechanical modifications (swelling, decrease of the hardness, increase of the fracture toughness, increase of the internal energy). A computational approach using classical molecular dynamics has been developed in parallel to clarify the correlations between the structural and mechanical property changes. This presentation will expose the main results from our simulations.

Computational method

An empirical pair potential of the Buckingham form has been developed recently to model atomic interactions [1]. Hence the interaction between any *(i,j)* pair of constituent atoms is modelled using the following relation:

$$\Phi(r_{ij}) = \frac{q_i q_j}{r_{ij}} + A_{ij}\exp\left(-\frac{r_{ij}}{\rho_{ij}}\right) - \frac{c_{ij}}{r_{ij}^6}$$

The adjustable parameters have been determined in order to reproduce structural and elastic properties of a set of SiO_2-B_2O_3-Na_2O glasses. A dependence between the ionic charges and the glassy compositions has been introduced in order to reproduce the Boron anomaly, i.e. the nonlinear evolution of the B coordination versus the R(=[Na_2O]/[B_2O_3]) and K(=[SiO_2]/[B_2O_3]) ratios. These potentials were then used to study fracture mechanisms and nanoindentation in pristine and irradiated simplified nuclear glasses.

Results

The three simulated glass compositions are the following:
SBN12 glass: 59.6%SiO_2 – 28.2%B_2O_3 – 12.2%Na_2O
SBN14 glass: 67.8%SiO_2 – 18.0%B_2O_3 – 14.2%Na_2O, this glass is characterized by the same molar ratios as the real nuclear glass
SBN55 glass: 55.3%SiO_2 – 14.7%B_2O_3 – 30.0%Na_2O

For each glassy composition, pristine structures (quenched from the liquid state with a rate equal to 5×10^{12}K/s) and disordered structures have been prepared. The disordered structures have been obtained after the heating of the pristine ones and their faster quench (with a rate equal to 10^{14}K/s) in order to reproduce the main radiation effects, i.e. depolymerization and increase of the internal disorder. The disordered glasses are considered as models of irradiated ones and their mechanical properties have been compared to those of the pristine glasses.

The fracture simulations have made possible to identify the different steps leading to the cracking of the materials [2]. Under the application of an external tensile stress, nano cavities are firstly formed. Then they begin to coalesce as the external stress increases forming increasingly large voids until the complete opening of the structure. During the fracture process, two different behaviors have been noticed: the Si and 4-coordinated B atoms on one hand accumulating stresses in their local environments, and, on the other hand, the 3-coordinated B and Na atoms adapting themselves much more easily to the local stress changes and then facilitating the viscous flow. This observation has made possible to propose an explanation of the origin of the fracture toughness increase in the simplified SBN14 glass. The radiation effects induce a decrease of the average B coordination, with a rise of the 3-coordinated B concentration. In consequence, the plasticity of the material increases, and the energy cost needed for the sample breakage is larger, hence the fracture toughness increase.

Concerning the nanoindentation simulations [3], a Vickers indenter in diamond has been pushed then removed from a simulation box and, using the loading – unloading curves, the hardness values have been quantified (Fig. 1). With the increase of the disorder, a systematic hardness decrease is observed [4] in the three glassy compositions that has been correlated to an increase of the 3-coordinated B and non-bridging oxygen concentrations.

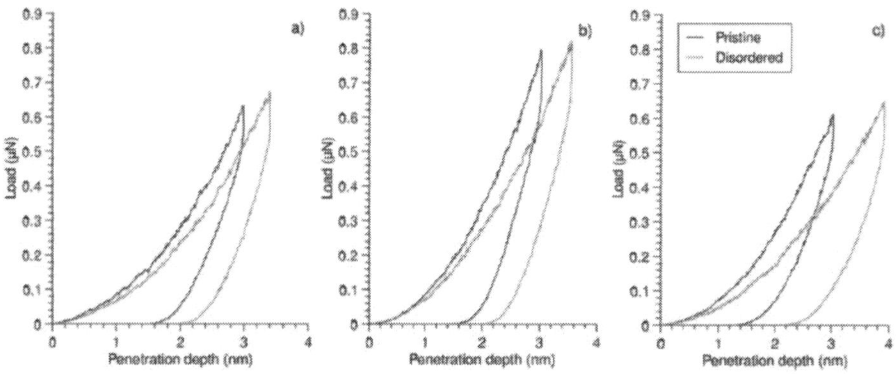

Fig. 1: Loading – unloading curves during nanoindentation simulated in the SBN12 (a), SBN14 (b) and SBN55 (c) glasses.

Taking into account the disordering and depolymerization effects, this study using the classical molecular dynamics method has therefore allowed reproducing changes of the mechanical properties analogous to what has been observed in simplified nuclear glasses irradiated externally by heavy ions. Thus, the structural origins of these changes have been better understood.

References

[1] L.-H. Kieu, J.-M. Delaye, L. Cormier, C. Stolz, “Development of empirical potentials for sodium borosilicate glass systems”, J. Non-Cryst. Solids 357 (2011) 3313.
[2] L.-H. Kieu, J.-M. Delaye, C. Stolz, “Modeling the effect of composition and thermal quenching on the fracture behavior of borosilicate glass”, J. Non-Cryst. Solids 358 (2012) 3268.
[3] D. A. Kilymis, J.-M. Delaye, “Deformation mechanisms during nanoindentation of sodium borosilicate glasses of nuclear interest”, J. Chem. Phys., 141 (2014) 014504.
[4] D. A. Kilymis, J.-M. Delaye, “Nanoindentation studies of simplified nuclear glasses using molecular dynamics”. J. Non-Cryst. Solids 401 (2014) 147.

EPJ Web of Conferences **115**, 02006 (2016)
DOI: 10.1051/epjconf/201611502006

MINOS
Centre of Excellence for Nuclear Materials

2nd Int. Workshop Irradiation of Nuclear Materials: Flux and Dose Effects
November 4-6, 2015, CEA – INSTN Cadarache, France

Simulation of C-Component Dislocation Loops by Ions and Protons Irradiation. A Tool to Understand the Breakaway Growth of Recrystallized Zr-Based Alloys

Jean Paul MARDON[1], Nesrine GHARBI[2], Thomas JOURDAN[3], Didier GILBON[4], Fabien ONIMUS[2], Xavier FEAUGEAS[5], Rosmarie HENGSTLER-EGER[6]

[1] AREVA NP, Fuel Business Unit (Lyon, France)
[2]CEA-DEN-DMN, Service de Recherches Métallurgiques Appliquées, SRMA (Saclay, France)
[3]CEA-DEN-DMN, Service de Recherches de Métallurgie Physique, SRMP (Saclay, France)
[4]CEA-DEN, Department of Nuclear Materials, DMN (Saclay, France)
[5]LaSI, University La Rochelle (La Rochelle, France)
[6]AREVA NP, GmBH (Erlangen, Germany)

During normal in-service operation, the PWR fuel assemblies exhibit a macroscopic elongation. This dimensional change is the result of three different phenomena: thermal creep, irradiation-creep and irradiation growth which occurs in the absence of any applied stress. This last phenomenon, referred as stress-free growth, is related to irradiation-induced microstructural changes taking place in the hexagonal close packed grains and corresponding to an elongation along the basal plane and a shortening along the c-axis, the volume remaining constant (Fig 1). Under neutron irradiation, recrystallized Zr-based alloys undergo stress-free growth which accelerates for high irradiation doses (Fig 2). This acceleration is correlated to specific c-component vacancy dislocation loops.

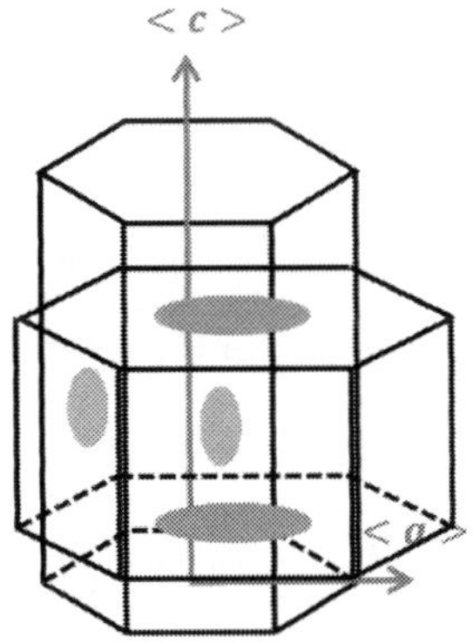

Fig. 1: Changes taking place in the hexagonal close packed grains [1]

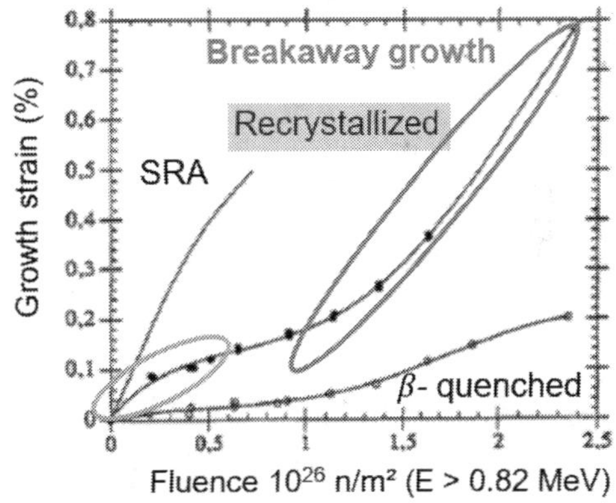

Fig. 2: Recrystallized Zr alloys breakaway growth [1]

Since these defects are responsible for the breakaway growth, which influences the performance of the fuel assembly, it is important to know the effect of various factors on their microstructure. The industrial feedback observed on some PWR fuel assemblies suggests that a macroscopic stress applied in the early life could affect the c-loop microstructure and therefore impact the subsequent stress-free growth. In addition, some feedbacks show that in-service hydrogen pick-up could also influence the fuel assembly radiation-induced deformation. In order to have a better understanding of this phenomenon, thorough analytical studies of c-loops nucleation and growth has been undertaken using charged particles irradiation. Protons and ions irradiation experiences have been carried out on various worldwide facilities at several temperatures combined with TEM exams [1, 2].

Two complementary studies were carried out to simulate neutron irradiation: in-situ tensile tests under 1MeV Kr^{2+} ion irradiation and bending experiments under 600keV Zr^{+} ion irradiation in order to investigate the impact of a macroscopic stress applied under irradiation on the evolution of c-loop microstructure [3].

In-situ Kr^{2+} ion irradiation was performed at high dose (up to 25dpa) on RXA Zircaloy-4 at 573K with the stress applied from the beginning of the irradiation. We observe in few grains a clear effect of stress on the c-loop linear density (Fig 3).

Zr+ ion irradiations were performed at 573K on RXA Zircaloy-4 and M5® and two loading histories were considered, using a four point bending device specifically designed to apply a tensile or a compressive stress under irradiation. During the first loading experiment, the external stress was applied at an early stage of irradiation. Transmission Electron Microscopy (TEM) observations showed that the initial applied stress has no effect on the incubation dose of c-loops. For the second loading experiment, the macroscopic stress was applied when c-loops are already created. Same trend, that the one observed for in-situ experience, but with a lower effect of stress on the c-loop linear density was observed on RXA Zircaloy-4 and M5® due to dispersion from grain to grain and a much larger statistic for this experiment (Fig 4) [1]. Nevertheless, qualitative agreement between two experiments is noticed.

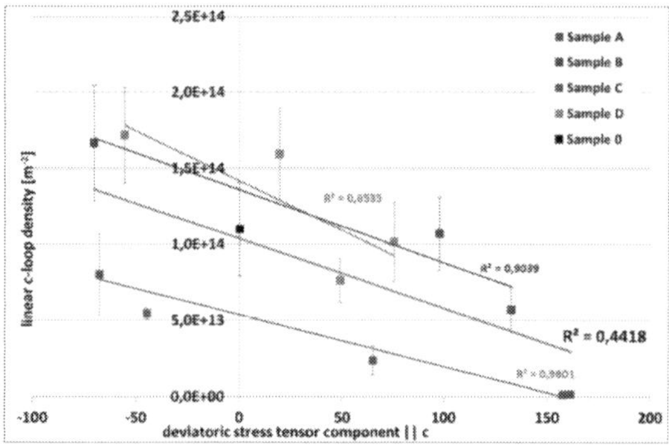

Fig. 3: Evolution of c-loops density for Zy4 (In-situ Kr^{2+} ion irradiation) [3]

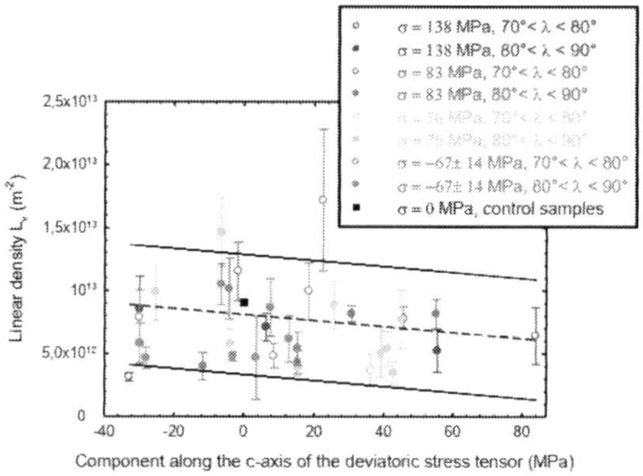

Fig. 4: Evolution of c-loops density for Zy4 (Bending Zr+ ion irradiations) [3]

Indeed, the c-loop linear density is lower when a tensile stress is applied parallel to the c-axis, which is in good agreement with the Stress Induced Preferential Absorption (SIPA) mechanism. Based on these results under stress level below the transverse yield strength and at 573K, the quantitative statistical study conducted on a large number of Zircaloy-4 and M5® grains with different orientations of the c-axis with respect to the loading direction, shows that an externally applied stress has a slight effect on the c-loop microstructure. A trend for the evolution of their densities with the component along the c-axis of the deviatoric stress tensor was obtained and is consistent with the SIPA mechanism, but seems to be not very significant considering the dispersion from grain to grain. The average loop size is weakly affected by the applied stress and the coupling between the axial irradiation creep and the accelerated irradiation growth appears to be low at the grain scale.

References

[1] N. Gharbi, F. Onimus, D. Gilbon, J.-P. Mardon, X. Feaugas, "Impact of an applied stress on c-component loops under Zr ion irradiation in recrystallized Zircaloy-4 and M5®", JNM-D-15-00312 (to be published).

[2] L. Tournadre, F. Onimus, T. Jourdan, J.-P. Mardon, X. Feaugeas, "Impact of hydrogen pick-up and applied stress on c-component loops: toward a better understanding of the radiation induced growth of recrystallized zirconium alloys", ASTM STP 1453 (2014), pp. 853-893.

[3] N. Gharbi, R. M. Hengstler-Eger, X. Feaugeas, D.Gilbon, P. B. Hoffmann, M. A.Kirk, J.-P.Mardon, F. Onimus, "The effect of applied stress on c-component dislocation loops in Zr-Based alloys", TMS 144th Annuel meeting and Exhibition, March 15-19, 2015, Orlando, USA.

EPJ Web of Conferences **115**, 03001 (2016)
DOI: 10.1051/epjconf/201611503001

MINOS
Centre of Excellence for Nuclear Materials

2nd Int. Workshop Irradiation of Nuclear Materials: Flux and Dose Efects
November 4-6, 2015, CEA – INSTN Cadarache, France

Microstructure Evolution in Fe and Fe-Cr Alloys with OKMC Methods

Maria J. CATURLA[1], Maria J. ALIAGA[1], Ignacio MARTIN-BRAGADO[2], Ignacio DOPICO[3], Lorenzo MALERBA[3], Mercedes HERNANDEZ-MAYORAL[4]

[1]Dept. Física Aplicada, Facultad de Ciencias, Fase II, University of Alicante (Alicante, Spain)
[2]IMDEA Materials Institute, Getafe (Madrid, Spain)
[3]SCK-CEN, Nuclear Materials Science Institute (Mol, Belgium)
[4]CIEMAT, Division of Materials (Madrid, Spain)

The type of damage observed under transmission electron microscopy (TEM) in b.c.c. Fe and FeCr alloys under irradiation is very dependent on the irradiation conditions. At high doses the damage consists mostly of ½ <111> and <100> loops, which are considered to be of interstitial type [1, 2]. Depending on the type of irradiation and the irradiation conditions, the ratio of one type of loop with respect to the other changes, with situations in which only one of the two types of loops are observed. The formation of <100> loops is still a controversial issue, with different models arising in the last few years. Experimentally, even the nature of these loops (vacancy or interstitial type) is still under debate under certain situations. Moreover, the type of irradiation (ion implantation vs. neutron irradiation) also changes the type and distribution of these defects.

In an attempt to shed some light into this complex system, we have developed models to study the damage of Fe and FeCr under irradiation. We combine results from density function theory (DFT), molecular dynamics (MD) and object kinetic Monte Carlo (OKMC) to study the production and evolution of the damage to time scales that can be directly compared to experiments. In this work we will discuss one particular issue: the differences between ion implantation and neutron irradiation. Currently, ion implantation is commonly used to obtain information about defect production and defect evolution, in an attempt to build models for neutron damage. We will point out the differences and commonalities between the two types of irradiation in the particular case of Fe, and those aspects that should be taken into consideration when building models for neutron irradiation from information obtained from ion implantation.

We will describe results of damage evolution in Fe using two different models for <100> loop formation: one where <100> loops are formed from the interaction between ½ <111> loops (based on simulations by Marian [3] and Terentyev [4]) and a second one where two populations of loops are formed directly in the collision cascade and evolve independently. This last model is based on the observation of stable immobile clusters from DFT calculations [5]. The population of loops formed as a function of dose for different irradiation conditions is explored and results compared to experiments. The validity of the models and the influence of the different parameters are discussed. In particular, we will study how the mobility of <111> interstitial loops changes both the total concentration of visible defects as well as the ratio of <111> vs. <100> loops and compare the results with experimental data available, as shown in the Fig. 1. In this case the experiments are from reference [1].

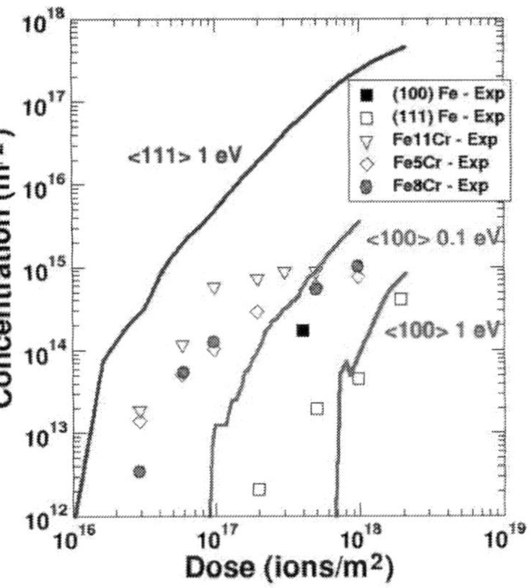

Figure 1: OKMC simulations (solid lines) compared to experiments from ref. [1]

Finally we will discuss how object kinetic Monte Carlo can be modified to study high concentrated alloys for the particular case of FeCr.

References

[1] Z. Yao, M. Hernández Mayoral, M. L. Jenkins, M. A. Kirk, *Heavy-ion irradiations of Fe and Fe-Cr model alloys Part 1: Damage evolution in thin-foils at lower doses*, Phil. Mag. 88 (2008) 2851.

[2] M.L. Jenkins, Z. Yao, M. Hernández-Mayoral, M.A. *Kirk, Dynamic observations of heavy-ion damage in Fe and Fe-Cr alloys,* J. of Nucl. Mat. 389 (2009) 197–202.

[3] J. Marian, B. D. Wirth and J. Manuel Perlado, *Mechanism of formation and growth of <100> interstitial loops in ferritic materials*, Phys. Rev. Lett. 88, 25 (2002).

[4] Haixuan Xu, Roger E. Stoller, Yuri N. Osetsky and Dmitry Terentyev, *Solving the puzzle of <100> interstitial loop formation in bcc iron*, Phys. Rev. Lett. 110, 265503 (2013).

[5] M.-C. Marinica, F. Willaime, J.-P. Crocombette, *Irradiation-induced formation of nanocrystallites with C15 laves phase structure in bcc iron*, Phys. Rev. Lett. 108, 025501 (2012).

EPJ Web of Conferences **115**, 03002 (2016)
DOI: 10.1051/epjconf/201611503002

MINOS
Centre of Excellence for Nuclear Materials

2nd **Int. Workshop** Irradiation of Nuclear Materials: Flux and Dose Effects
November 4-6, 2015, CEA – INSTN Cadarache, France

Atomistic Modelling of Segregation and Precipitation in Fe-Cr Alloys under Irradiation

Frédéric SOISSON[1], Chu Chun FU[1], Thomas JOURDAN[1], Maylise NASTAR[1], Oriane SENNINGER[1], Enrique MARTINEZ[2], Yves BRÉCHET[3]

[1]CEA-DEN-DMN, Service de Recherches de Métallurgie Physique, SRMP (Saclay, France)
[2]Los Alamos National Laboratory, Materials Science in Radiation Dynamics Group (Los Alamos, USA)
SIMAP, INP Grenoble, CNRS UJF (Saint Martin d'Hères, France)

Iron-chromium alloys are the basis for ferritic and ferritic-martensitic steels that will be used in future fission (generation IV) and fusion nuclear reactors. With Cr content between typically 8 to 12% [1], or even 14% in the matrix of some oxide dispersion-strengthened steels [2], one can expect the precipitation of a Cr-rich α' phase that can be strongly accelerated under irradiation, due to point defect supersaturation. This precipitation can cause hardening and embrittlement. Radiation-induced segregation (RIS) is another important technological problem. It can lead to a Cr depletion at grain boundaries and therefore to a loss of corrosion resistance and again, to embrittlement. RIS in austenitic steels is well known and presents almost systematic trends: depletions of Cr and enrichments of Ni at grain boundaries. In ferritic steels, the experimental situation is far from being so clear: depletions and enrichments of Cr have been observed, without clear correlation with the irradiation conditions and the materials properties [3].

Segregation and precipitation occur by formation, migration and elimination of point defects (vacancies and self-interstitials). We present a model that includes these mechanisms, and combines ab initio and atomistic kinetic Monte Carlo (AKMC) simulations. Migration barriers and jump frequencies are computed by using effective pair interactions with concentration and temperature dependences. Thanks to these dependences, the simulations are able to well reproduce the thermodynamic [4] and diffusion [5] properties of dilute and concentrated alloys, including the effects of the magnetic configurations and magnetic transitions, which are especially important in Fe-Cr alloys. They also give kinetics pf precipitation during isothermal annealing in good agreement with experimental studies [5].

Simulations of α' precipitation under irradiation, predict that the kinetics can be considerably accelerated. For example, irradiations of alloys with 15 to 18%Cr at 290°C and a dose rate of 3.4×10^{-7} dpa.s^{-1} produce a point defect supersaturations and therefore, an acceleration, of approximately 6-7 orders of magnitudes. To get a more precise estimation of the evolution of sink densities and point defect concentrations, cluster dynamics [6] have been used: the physical time scale of the Monte Carlo simulations is rescaled accordingly and one gets precipitation kinetics in good agreement with the few available experimental results [7,8]. The possible effect of ballistic mixing occurring in displacement cascades and the role of carbon, which is known to strongly interact with the vacancies, will be discussed.

RIS results from the elimination of excess vacancies (V) and self-interstitials (I) on sinks such as dislocations, grain boundaries, or free surfaces. Permanent fluxes of point defects towards the sinks are then sustained, i.e. permanent fluxes of chemical species, leading to a modification of the local composition near the sinks. RIS can be analyzed properly in the framework of the thermodynamics of irreversible processes, that gives the flux of a specie α (atom or defect) as a function of the gradients of chemical potentials μ_β : $J_\alpha = -\sum L_{\alpha\beta}\nabla\mu_\beta$. The tendencies to solute enrichment or depletion near the sinks are controlled by the Onsager coefficients, $L_{\alpha\beta}$. In Fe-Cr alloys, one must observe a depletion of Cr near the sinks if $L_{CrV}/L_{FeV} > L_{CrI}/L_{FeI}$, an enrichment of Cr in the opposite case. Onsager coefficients can be directly estimated in Monte Carlo simulations (Fig. 1a). Simulations have been also performed to model directly the evolution of point defects and solute distributions near a grain boundary (Fig. 1b). As expected from the analysis of the Onsager coefficients, at 650 K the interstitial contribution dominates and one observes an enrichment of Cr. On the contrary at 950K vacancy contribution is dominant and Cr depletion occur [9]. We will compare these trends with experimental observations of RIS. Finally, the AKMC enables to study the interaction between segregation and precipitation [9]: typical cases will be presented.

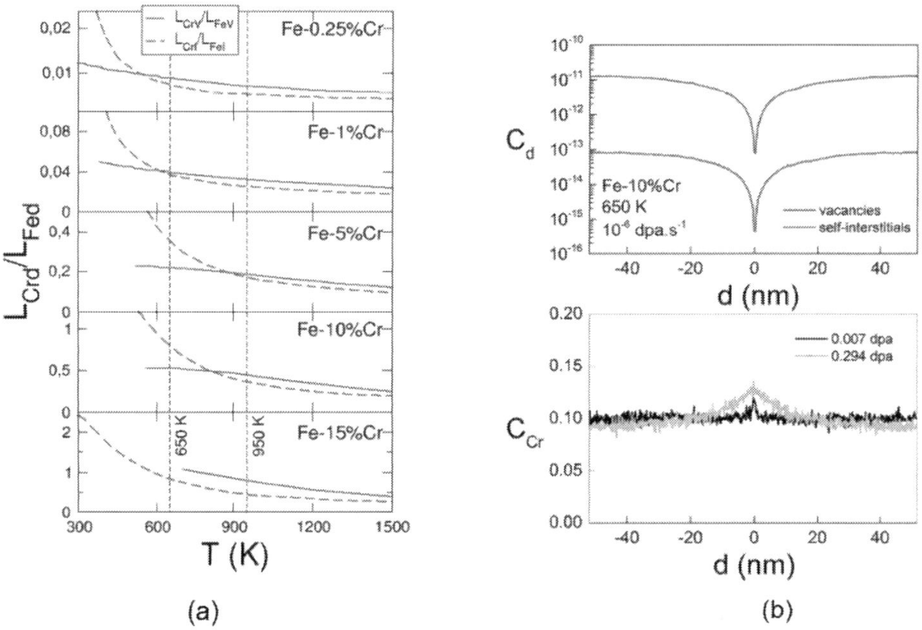

Fig. 1: (a) Evolution of the ratios of Onsager coefficients with the temperature, in Fe-Cr alloys of different compositions, estimated by Monte Carlo simulations, (b) Monte Carlo simulation of radiation induced segregation of Cr near a grain boundary (steady-state concentration profiles of point defects and concentration profiles of Cr at two successive doses).

References

[1] R.L. Klueh, A.T. Nelson, *Ferritic/martensitic steels for next-generation reactors*, J. Nucl. Mater. 371 (2007) 37.
[2] J.-L. Boutard et al, *Oxide dispersion strengthened ferritic steels: a basic research joint program in France*, J. Nucl. Mater. 455 (2014) 605.
[3] Z. Lu et al, *Irradiation-induced grain boundary chromium microchemistry in high alloy ferritic steels*, Scripta Mater. 58 (2008) 878
[4] M. Levesque, E. Martínez, C.-C. Fu, M. Nastar and F. Soisson, *Simple concentration-dependent pair interaction model for large-scale simulations of Fe-Cr alloys*, Phys. Rev. B 84, 184205 (2011)
[5] O. Senninger, E. Martinez, F. Soisson, M. Nastar and Y. Bréchet, *Atomistic simulations of decomposition kinetics in Fe-Cr alloys: influence of the magnetism*, Acta Mater. 73, 97–106 (2014).
[6] E. Meslin et al, *Cluster-dynamics modelling of defects in α-iron under cascade damage conditions*, J. Nucl. Mater. 382 (2008) 190.
[7] M. Bachhav, G. R. Odette, and E. A. Marquis, *α' precipitation in neutron-irradiated Fe-Cr alloys*, Scripta Mater. 74 (2014) 48.
[8] V. Kuksenko, C. Pareige and P. Pareige, *Cr precipitation in neutron irradiated industrial purity Fe–Cr model alloys*, J. Nucl. Mater. 432 (2013) 160.
[9] O. Senninger et. al., *Modeling radiation induced segregation in Fe-Cr alloys*, submitted to Acta Mater.

EPJ Web of Conferences **115**, 03003 (2016)
DOI: 10.1051/epjconf/201611503003

MINOS Centre of Excellence for Nuclear Materials
2nd Int. Workshop Irradiation of Nuclear Materials: Flux and Dose Effects
November 4-6, 2015, CEA – INSTN Cadarache, France

Study of Rare Gases Behavior in Uranium Dioxide

Gaëlle CARLOT[1], Catherine SABATHIER[1], Serge MAILLARD[1], Amélie MICHEL[1], Guillaume MARTIN[1], Eric GILABERT[2], Franck FORTUNA[3], Philippe GARCIA[4]

[1]CEA-DEN-DEC, Service d'Etudes et de Simulation du Comportement des Combustibles, SESC (Cadarache, France)
[2]Centre d'Etudes Nucléaires de Bordeaux Gradignan, CNRS/IN2P3 (Gradignan, France)
[3] Centre de Sciences Nucléaires et de Sciences de la Matière, CNRS/IN2P3 (Orsay, France)
[4]CEA-DEN-DEC, Service Fabrication, Elaboration et Reconditionnement des Combustibles, SFER (Cadarache, France)

The behaviour of fission products in uranium dioxide has been a focus of considerable experimental and theoretical attention in the nuclear industry. Xenon and krypton have attracted particular attention due to their high fission yields. During in-reactor irradiation of the nuclear fuel, rare gases are subject to several phenomena: diffusion and precipitation. These phenomena can have adverse consequences on the fuel physical and chemical properties and its in-reactor behavior. When released from the material they induce an increase in the fuel rod pressure detrimental to its integrity. A greater understanding of the mechanisms that underpin rare gas diffusion and precipitation should enable the development of models with a greater predictive capability and also make it possible to optimize fuel microstructures so as to increase the fuel element discharge burnup. The development of models for describing oxide behavior is hindered by the lack of experimental data available. The aim is to contribute to the understanding of the prevailing phenomena and generate sets of fundamental data necessary for modelling rare gas behaviour. The purpose of this work is to better understand the behavior of fission gases by identifying diffusion and bubble nucleation. To do this, studies involving separate effects have been established coupling ion irradiations/implantations with fine characterizations (Thermal Desorption Spectrometry and Transmission electron microscopy).

In order to analyse the mechanisms involved in rare gases diffusion, we carried out Thermal Desorption Spectrometry (TDS) measurements on the PIAGARA (Plateforme Interdisciplinaire pour l'Analyse des GAz Rares en Aquitaine) platform at CENBG laboratory in Bordeaux on UO_2 samples implanted at low fluence [Fig. 1]. The experiment involves heating a sample in a small furnace to induce rare gases movement and release and using mass spectrometry to monitor the amount of rare gases released as a function of time. Interpretation of the release experiments has enabled us to determine xenon and krypton diffusion coefficients in uranium dioxide

Fig. 1: Plateform PIAGARA at CENBG laboratory

Transmission electron microscopy (TEM) observations of UO_2 polycrystals irradiated in situ with 4 MeV Au ions were performed at room temperature (RT) to better understand the mechanisms of cavity and ultimately fission products nucleation in UO_2. Experiments were carried out at the JANNuS Orsay facility that enables in situ ion irradiations inside the microscope to be carried out. The majority of 4 MeV gold ions were transmitted through the thin foil, and the induced radiation defects were investigated by TEM. Observations showed that nano-void formation occurs at ambient temperature in UO_2 thin foils irradiated with energetic heavy ions under an essentially nuclear energy loss regime. The diameter and density of nano-objects were measured as a function of the gold irradiation dose at RT [Fig. 2]. A previous study [1] has also revealed a similar nano-object population after a Xe implantation performed at 390 keV at 870 K. The nano-object density was modelled using simple concepts derived from Classical Molecular Dynamics simulations. The results are in good agreement, which suggests a mechanism of heterogeneous nucleation induced by energetic cascade overlaps. This indicates that nano-void formation mechanism is controlled by radiation damage. Such nanovoids are likely to act as sinks for mobile fission products during reactor operation.

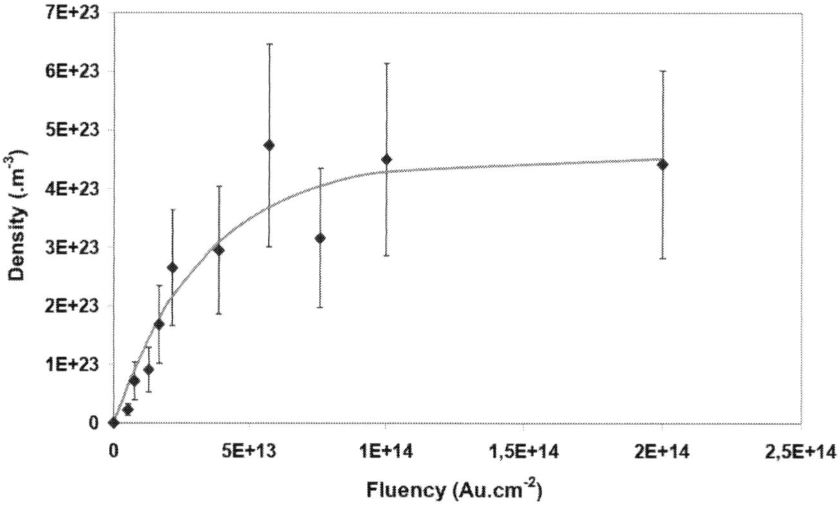

Fig. 2: Nano-void density evolution with Au irradiation at RT from TEM observation.

References

[1] A. Michel, C. Sabathier, G. Carlot, O. Kaïtasov, S. Bouffard, P. Garcia, C. Valot, Nuclear Instruments and Methods in Physics Research Section B: Beam Interactions with Materials and Atoms, 272 (2012) 218.

EPJ Web of Conferences **115**, 03004 (2016)
DOI: 10.1051/epjconf/201611503004

MINOS Centre of Excellence for Nuclear Materials
2nd **Int. Workshop** Irradiation of Nuclear Materials: Flux and Dose Effects
November 4-6, 2015, CEA – INSTN Cadarache, France

Positron Annihilation Spectroscopy to Characterize Irradiation Induced Vacancy Type Defects in Materials for Nuclear Fission and Fusion

Marie-France BARTHE[1]

[1]CEMHTI, CNRS, Université d'Orléans (Orléans, FRANCE)

In nuclear reactors, materials are submitted to irradiations and modification of their macroscopic properties such as swelling or hardening is observed... It is of first importance to understand the origin of these evolutions and how damage at the atomic scale such as vacancy and interstitials type defects can interact with each other or with solutes and impurities to make the microstructure evolved. So the properties of defects have to be determined.

Positron annihilation spectroscopy (***PAS***) is a well-established technique to characterize materials [1]. Due to its positive charge, positron is sensitive to the local variations of the Coulomb potential in solids. As the anti-particle of the electron, both particles can annihilate leading to the emission of gamma rays with energy depending on the momentum of the electron positron annihilated pair. The detection of these gamma rays arises original information on the defects because positron can be trapped and annihilate in localized state such as vacancies. Two different and complementary annihilation characteristics can be measured. Firstly Positron Annihilation Lifetime Spectrometer (***PALS***) allows to measure the *positron lifetime* which depends on the local electron density at the annihilation site. The second characteristic is the momentum distribution of the electron positron annihilated pairs and is obtained by measuring the energy spectrum of the gamma annihilation rays using a Doppler Broadening Spectrometer (***DBS***). Both characteristics give the signature of the defects and allow determining some of their properties such as their nature, concentration, chemical environment... By using slow positron accelerators which produce monokinetic positrons beams with energy varying from 0.1 to 30 keV it is possible to study the depth profile of defects in thin layers from 0.1 to about 5 µm depending on the density of materials with a resolution of the order of 0.1xdepth. ^{22}Na source based beams [2] or user facilities taking advantages of high positrons flux available around nuclear reactor (FRMII, Garching [3]) or LINAC(as in AIST in Japan [4]) are now available to study defects in materials. PAS has numerous advantages among them, it is non-destructive and can be used for conducting as well as insulating materials crystalline or amorphous. This technique is especially useful to probe vacancy defects in metals for which it is one of the only direct characterization. The annihilation characteristics can be predicted by first principle calculations and these data when available can help to identify defects. Sensitive to the single vacancy, PAS can allow to determine the size of the vacancy clusters up to a maximum of 1 nm approximately in the concentration range from 10^{15} to a few 10^{18} cm^{-3}.

In nuclear science this technique is very powerful to characterize not only damage induced by irradiation but also to determine some fundamental properties of defects that are required for modeling of the microstructural evolution of materials such as formation, migration, agglomeration, interaction with solutes, impurities or transmutation products such as Helium or Hydrogen. PAS has already demonstrated its major role in the study of vacancy defects and material related properties. For example the results obtained by Vehanen et al [6] have had an outmost impact on the knowledge of the role of carbon on vacancy behavior in Fe. They have demonstrated the trapping of vacancy in Carbon Vacancy complexes and the non-localization of C atoms in the center of the vacancy. This attractive interaction and localization has been confirmed by ab initio calculations [7]. It has also to be underlined that positrons can be trapped in precipitates if its affinity for the material of precipitate is higher than the one of the matrix. Nagai et al have demonstrated the complementarity of positron techniques with Atomic Probe Tomography and have recently underlined the effects of matrix damage probed by PAS on the hardening [5].

In this work we will report different results on the determination of the properties of vacancy defects in W and UO_2 using separate effects irradiation experiments coupled to PAS experimental measurements and theoretical data. Recently the self-consistent two components Density Functional Theory has been implemented in PAW approximation and has allowed circumcising the nature of the detected vacancy defects in UO_2 irradiated with electrons or alpha particles [8]. The comparison of the experimental positron lifetime with calculated values show that the Schottky defects made of one uranium vacancy associated to 2 oxygen vacancies seem to be the most detected defects. Due to its properties tungsten has been chosen to cover part of the divertor in ITER and is envisaged for first walls in next tokamak generation such as DEMO. In order to better understand its swelling and embrittlement observed in special irradiation conditions, we use PAS to study the vacancy defects properties in this material. We have first shown the migration and agglomeration of single vacancy in the temperature range from 473 to 600K [9] and its interaction with helium [10]. More recently we have observed the formation of vacancy clusters in self ion irradiated W and the effect of the temperature on the size of these clusters. Today, the development of positron microbeams is very promising. It has already allowed to show the creation of vacancies in region close to the fracture zone of H charging stainless steel [11].

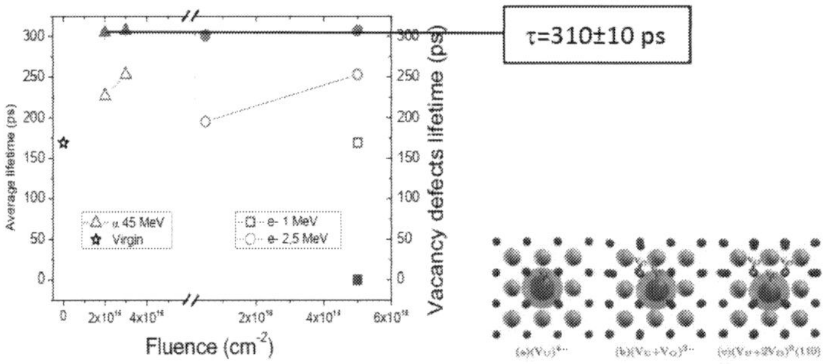

Fig. 1 : Positron Lifetime in electron and alpha irradiated UO_2 and positron isodensities in Uranium vacancy and complexes with oxygen vacancy as calculated in self-consistent 2 components Density Fonctionnal Theory [8]

References

[1] M.-F. Barthe et al., Techniques de l'Ingénieur P3(P145) 2610 (2003).
[2] P Desgardin et al.," Slow positron beam facility in Orleans", Materials science Forum 363, 523 (2001).
[3] https://www.frm2.tum.de/en/the-neutron-source/secondary-sources-for-neutrons-and-positrons/positron-source/
[4] https://unit.aist.go.jp/rima/xr-pos/group_detail/facilities/facilities.html
[5] A. Kuramoto et al., "Microstructural changes in a Russian-type reactor weld material after neutron irradiation, post-irradiation annealing and re-irradiation studied by atom probe tomography and positron annihilation spectroscopy", Acta Materialia 61, 5236 (2013).
[6] A. Vehanen et al., "Vacancies and carbon impurities in α-Fe- electron irradiation", Phys. Rev. B 25, 762, (1982).
[7] C. Barouh et al.," Interaction between vacancies and interstitial solutes (C, N, and O) in α -Fe: From electronic structure to thermodynamics", Phys. Rev. B 90, 054112 (2014).
[8] J. Wiktor et al., "Coupled experimental and DFT plus U investigation of positron lifetimes in UO_2", Phys. Rev. B 90, 184101 (2014).
[9] P. E. Lhuillier et al., "Positron annihilation studies on the nature and thermal behaviour of irradiation induced defects in tungsten", Physica Status Solidi C 6, 2329 (2009).
[10] P. E. Lhuillier et al., "Helium retention and early stages of helium-vacancy complexes formation in low energy helium-implanted tungsten", JNM 433, 305 (2013).
[11] N. Oshima et al., Rapid three-dimensional imaging of defect distributions using a high-intensity positron microbeam", Appl. Phys. Lett. 94, 194104 (2009).

EPJ Web of Conferences **115**, 03005 (2016)
DOI: 10.1051/epjconf/201611503005

MINOS Centre of Excellence for Nuclear Materials

2nd Int. Workshop Irradiation of Nuclear Materials: Flux and Dose Effects
November 4-6, 2015, CEA – INSTN Cadarache, France

α Self-irradiation Effects on Structural Properties of $(U,Am)O_{2\pm\delta}$ Materials

Florent LEBRETON[1], Denis HORLAIT[2], Damien PRIEUR[3], Philippe MARTIN[1], Jean-Yves COLLE[3], Arne JANSSEN[3], Thierry WISS[3], Andreas SCHEINOST[4], Thibaud DELAHAYE[5]

[1]CEA-DEN-DTEC, Service d'Études des Combustibles et matériaux à base d'Actinides, SECA, (Marcoule, France)
[2]CEA-DEN-DTEC, Service de Développement des Technologies du Cycle, SDTC, (Marcoule, France)
[3]European Commission, Joint Research Centre, Institute for Transuranium Elements (Karlsruhe, Germany)
[4]Helmholtz Zentrum Dresden Rossendorf (HZDR), Institute of Resource Ecology (Dresden, Germany)
[5]CEA-DEN-DRCP, Service d'Étude du Recyclage des combustibles et d'Analyses, SERA, (Marcoule, France)

Am isotopes are amongst the largest contributors to the radiotoxicity and heat load of spent nuclear fuels, especially in scenarios including plutonium recycling in MOX fuels. Transmutation in heterogeneous mode is a promising option to lower Am inventory in the ultimate nuclear waste, thus decreasing the ecological impact of deep-geological repository sites. It consists in the irradiation Am-bearing blankets (AmBB, i.e., $U_{1-x}Am_xO_{2\pm\delta}$ with x values close to 10-15 at.%) in the periphery of a fast neutron reactor core. In this context, studies are being performed to assess the feasibility and safety of AmBB transmutation, notably focusing on innovative fabrication processes [1–3], AmBB behavior under irradiation [4] and $U_{1-x}Am_xO_{2\pm\delta}$ properties [5]. ^{241}Am being a strong α emitter (activity of 130 MBq), $U_{1-x}Am_xO_{2\pm\delta}$ behavior under α self-irradiation is also monitored [6–8].

One of the main effects of α radiation in crystallized materials is a structural swelling. $U_{1-x}Am_xO_{2\pm\delta}$ compounds with different Am/(U+Am) ratios (7.5 ≤ x ≤ 50 at.%) were monitored by X-ray diffraction (XRD) after a heat treatment during several months to observe their lattice parameter evolution with time. The results obtained first evidenced that the fluorite-type structure (*Fm*-3*m*) of the compounds remains stable all along the study. During the first days of storage, all samples underwent an oxidation, observed for the lowest Am/(U+Am) ratios (x ≤ 15 at.%) through a decrease of the lattice parameter or, for the other (30, 40 and 50 at.%), through an isostructural transition to a phase with a smaller lattice parameter (Fig. 1 left). After this phenomenon, an increase of the lattice parameter was observed for all the compositions, evidencing an asymptotic behavior leading to saturation after a few months. This evolution was fitted with the relation describing the structural swelling under α irradiation: $(a - a_0)/a_0 = A(1 - e^{-Bt})$, with a, the lattice parameter at a time t, a_0 the initial lattice parameter, A the relative swelling at saturation and, B, a kinetics constant. Satisfactory fitting results were obtained for all samples (with R^2-factors superior to 99.5 %, see Fig 1 right). A parameters obtained indicate that all compositions present lattice swellings at saturation ranging from 0.26 to 0.28 %, with a small decrease with the Am content. The swelling kinetics logically increases with the Am content, but similar kinetics are obtained for all samples after normalization by the sample α activity. Additional experimental data was acquired using X-ray absorption spectroscopy (XAS), which confirmed the stability of the fluorite structure and that the different interatomic distances (actinide-O and actinide-actinide) present an evolution similar to that of the lattice parameter.

The effects of α self-irradiation in terms of structural disorder were also studied based on XRD and XAS data. Microstrain determined from XRD by the Williamson-Hall method did not exhibit any significant evolution with time (whereas it increases with the Am content, in line with previous results [5]). The Debye-Waller factors extracted from XAS however behaves similarly to the lattice parameter, *i.e.*, increase followed by saturation, even though fewer points were available to give a precise trend. The reason for this discrepancy is probably that EXAFS gives information on the whole sample, including low-ordered or amorphous domains, whereas XRD results only concern crystallized domains. The Debye-Waller factor increase with time thus appears to be related to an accumulation of the α-induced defects in domains with no (or low) long range order, such as the vicinity of grain boundaries. This phenomenon of defect concentration then favors their recovery by recombination and is thus responsible for the saturation in both lattice parameter and defect increase with time.

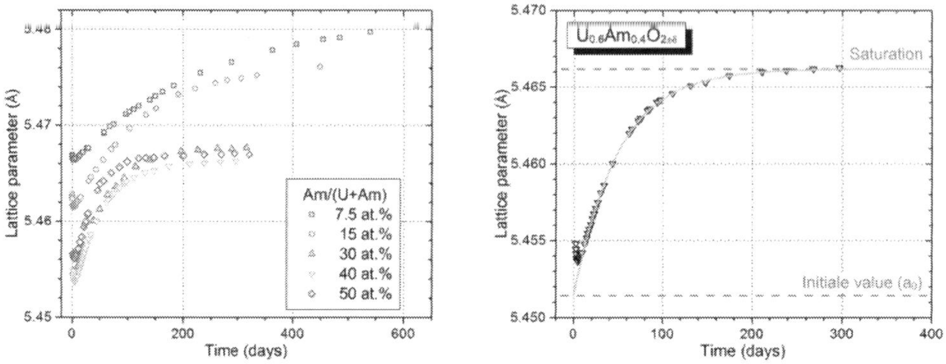

Fig. 1: Evolution of $U_{1-x}Am_xO_2$ lattice parameters with time (left), and example of fit for the $U_{0.6}Am_{0.4}O_{2\pm\delta}$ compound (right).

Studies were also dedicated to the behavior of He atoms produced during α decay of ^{241}Am. A 5-year-old $U_{0.85}Am_{0.15}O_{2\pm\delta}$ sample was heat-treated up to 1500 K in a Knudsen cell coupled with a mass spectrometer (MS) allowing the quantification of released He. Transmission electron microscopy (TEM) observations were also performed on the sample before and after thermal treatment. Results show that a first He release occurs around 1000 K, followed by a main one around 1350 K, as shown in Fig. 2. The total quantity of He released during this heat treatment was estimated to represent about 75 % of the He produced by α disintegrations during the sample storage. TEM micrographs show that only subnanometric cavities (He bubbles) are observed before thermal treatment, whereas larger cavities are observed after, with diameters typically ranging between 2 and 5 nm. These results thus evidenced a coalescence of He trapped in the sample during the thermal treatment leading both to its release and to bubble growth.

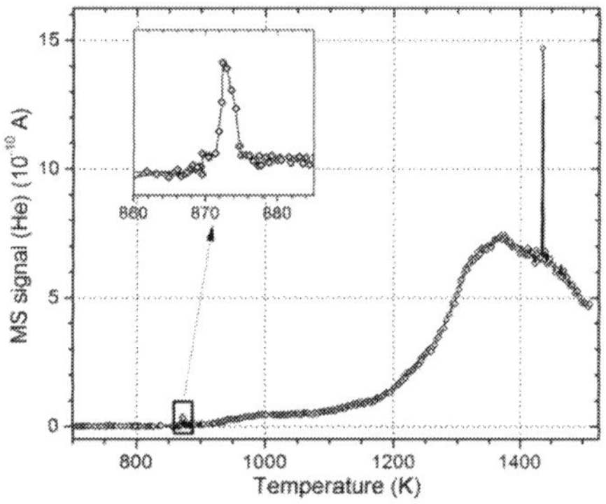

Fig. 2: He release rate from a 5-year-old $U_{0.85}Am_{0.15}O_{2\pm\delta}$ sample measured by MS.

References

[1] T. Delahaye, F. Lebreton, *et al.*, "Application of the UMACS process to highly dense $U_{1-x}Am_xO_{2\pm\delta}$ MABB fuel fabrication for the DIAMINO irradiation", J. Nucl. Mater. 432 (2013) 305–312.

[2] D. Horlait, F. Lebreton, *et al.*, "Americium-based oxides: Dense pellet fabrication from co-converted oxalates", J. Nucl. Mater. 444 (2014) 181–185.

[3] E. Remy, S. Picart, *et al.*, "Fabrication of uranium–americium mixed oxide pellet from microsphere precursors: Application of CRMP process", J. Nucl. Mater. 453 (2014) 214–219.

[4] S. Bejaoui, "Americium-Bearing Blanket Separate-effect Experiments: MARIOS and DIAMINO Irradiations" in:, Proc. Glob. 2011, Makuhari, Japan, 2011.

[5] D. Prieur, P.M. Martin, *et al.*, "Local Structure and Charge Distribution in Mixed Uranium–Americium Oxides: Effects of Oxygen Potential and Am Content", Inorg. Chem. 50 (2011) 12437–12445.

[6] D. Prieur, P.M. Martin, *et al.*, "Alpha self-irradiation effect on the local structure of the $U_{0.85}Am_{0.15}O_{2\pm x}$ solid solution", J. Solid State Chem. 194 (2012) 206–211.

[7] D. Horlait, F. Lebreton, *et al.*, "XRD Monitoring of α Self-Irradiation in Uranium–Americium Mixed Oxides", Inorg. Chem. 52 (2013) 14196–14204.

[8] F. Lebreton, P.M. Martin, *et al.*, "New Insight into Self-Irradiation Effects on Local and Long-Range Structure of Uranium–Americium Mixed Oxides (through XAS and XRD)", Inorg. Chem. 53 (2014) 9531–9540.

EPJ Web of Conferences **115**, 04001 (2016)
DOI: 10.1051/epjconf/201611504001

MINOS
Centre of Excellence for Nuclear Materials

2nd Int. Workshop Irradiation of Nuclear Materials: Flux and Dose Effect
November 4-6, 2015, CEA – INSTN Cadarache, France

Dislocation Grain Boundary Interactions in Irradiated Metals

Ian M. ROBERTSON[1], Bai CUI[2], Mo-Rigen HE[1]

[1]Dept. of Materials Science and Engineering, University of Wisconsin-Madison, (Madison, USA)
[2]Dept. of Mechanical and Materials Engineering, University of Nebraska-Lincoln (Lincoln, USA)

The behavior of dislocations in unirradiated and ion-irradiated austenitic stainless steels has been observed directly by performing straining experiments in-situ in the transmission electron microscope. This has enabled direct observation of how dislocations interact with each other, with the irradiation produced obstacles and with grain boundaries. These experiments have provided insight as to the mechanisms by which dislocations annihilate the irradiation produced defects to form a channel as well as determination of the conditions that control slip transfer across a grain boundary and how it is impacted by the presence of the irradiation produced defects.

The mobility of dislocations across a grain are fundamentally different in unirradiated and irradiated austenitic stainless steels. In the unirradiated case, pre-existing as well as newly generated dislocations move rapidly, smoothly and as a continuous line elements through the grain. In contrast in irradiated metals, pre-existing dislocations or ones created prior to irradiation are rendered immobile by the irradiation. This locking of dislocations by the irradiation was demonstrated directly by straining an electron transparent sample in-situ in the TEM to generate mobile dislocations, irradiating the sample, continuing the straining, and observing the response of the dislocations [1]. The mobility of these dislocations was very limited. The deformation was carried by generating new dislocations from either grain boundary sources or small cracks generated during sample preparation. This observation demonstrates that pre-existing dislocations are not responsible for the formation of dislocation channels in deformed irradiated metals. A second difference is the movement of dislocations through the field of obstacles, it now occurs in an irregular, jerky and discontinuous manner. The initial dislocation velocity was low as the channel with reduced obstacle density was formed. The velocity increased with the passage of additional dislocations, but never become a simple easy glide path with dislocations moving as in the unirradiated material. Instead with the passage of additional dislocations cross-slip events against barriers invisible in the TEM became more numerous. This lead to the formation of long straight dislocations along the boundary between the irradiated matrix and the dislocation channel [2].

On intersecting a grain boundary dislocations can be accommodated within it. This occurred in both the unirradiated and irradiated materials. These accommodated dislocations may retain the matrix Burgers vector or dissociate into grain boundary type dislocations. These dislocations can be either sessile or glissile within the grain boundary. If the dislocations are glissile, the build-up of stress is distributed along the grain boundary and the location on the grain boundary from which dislocations are emitted is unlikely to be coincident with that of dislocation impact. Importantly, many dislocations were observed to be accommodated within the grain boundary and a pile-up of dislocations developed against it before slip was transmitted across it [3]. A consequence of this is that the dislocations emitted from the grain boundary source are in general not the same as the ones that intersected it. That is, new dislocations are created and emitted from within the grain boundary. The exception is for screw dislocations intersecting a twin boundary where the line of intersection of the two slip planes in the grain boundary is collinear with the line direction, Burgers vector, of the dislocation in both grains. The process of accommodating the incoming dislocations in and the emission of dislocations from the grain boundary disrupts the internal structure of the grain boundary and for high levels of transfer can cause a displacement of the grain boundary.

The propagation of dislocations through grain boundaries in unirradiated FCC metals has been shown to be governed by the magnitude of the Burgers vector of the grain boundary dislocation generated by the process of transmission of slip across the grain boundary [3, 4]. The local resolved shear stress on the selected slip system must be sufficient to propagate the dislocations away from the grain boundary source into the grain. In irradiated FCC metals, the slip system activated to transmit slip is still governed by the magnitude of the grain boundary dislocation generated, but now the local resolved shear stress acting on the activated slip system must be of sufficient magnitude to propel the emitted dislocations through the irradiation hardened grain [5, 6]. Evidence supporting this claim was found in the in-situ TEM experiments, as the magnitude of the local shear stress, through the addition of dislocations to the pile-up, had to be increased before the dislocations could propagate away from the grain boundary. Effectively, the irradiated matrix transitions the rate limiting step for slip transfer across a grain boundary from one of dislocation nucleation to dislocation propagation. If this local shear stress is insufficient the dislocations are not emitted from the grain boundary and slip is not transferred across it.

An interesting question to ask is what happens in situations in which the magnitude of the local resolved shear stress never becomes sufficient for slip transfer to occur. In such cases, one possibility is for the grain boundary to relieve the accumulated stress by disrupting the grain boundary structure by nucleating and propagating a crack along it. Although such events are rare they are possible and have been observed in both in-situ TEM straining experiments as well as in macroscale mechanical property tests. This observation has important implications for breaking the protective oxide and exposing the base metal to the environment, which has implications for irradiation assisted stress corrosion cracking. Examples of the different interactions dislocations can have with a grain boundary are illustrated in Figure 1, which shows dislocation accommodation in the grain boundary with no slip transmission, Fig. 1a; dislocation slip along the grain boundary, Fig. 1b; slip transmission involving one incoming system generating two outgoing systems, Fig. 1c; displacement of the grain boundary following extensive slip transmission, Fig. 1d, and crack nucleation and propagation along a grain boundary, Fig. 1e.

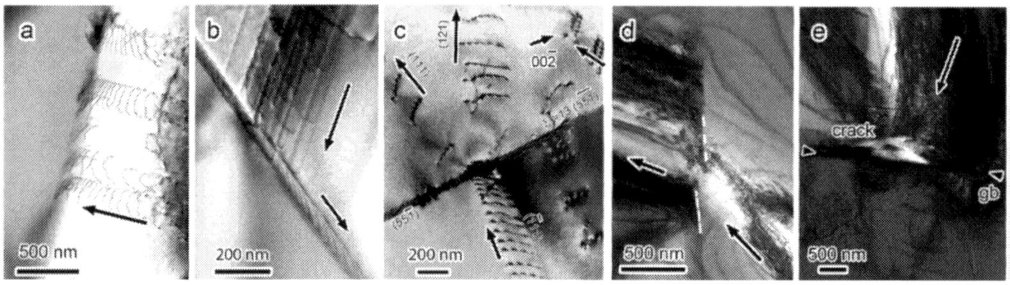

Fig. 1: Dislocation interactions with grain boundaries: a. accommodation, b. slip along grain boundary, c. slip transmission, d. slip transmission showing change in boundary structure and e. displacement of grain boundary due to slip transmission. a-d unirradiated and e. irradiated.

References

[1] Briceño M, Fenske J, Dadfarnia M, Sofronis P, Robertson IM. Effect of ion irradiation-produced defects on the mobility of dislocations in 304 stainless steel. J. Nucl. Mater. 2011; 409:18.
[2] Kacher J, Liu GS, Robertson IM. In situ and tomographic observations of defect free channel formation in ion irradiated stainless steels. Micron 2012; 43:1099.
[3] Kacher J, Eftink B, Cui B, Robertson IM. Dislocation interactions with grain boundary interactions. Current Opin. Solid State and Mater. Sci. 2014; 18 227.
[4] Spearot DE, Sangid MD. Insights on slip transmission at grain boundaries from atomistic simulations. Current Opin. Solid State and Mater. Sci. 2014; 18:188.
[5] Cui B, McMurtrey MD, Was GS, Robertson IM. Micromechanistic origin of irradiation-assisted stress corrosion cracking. Phil. Mag. 2014:1.
[6] Cui B, Kacher J, McMurtrey M, Was G, Robertson IM. Influence of irradiation damage on slip transfer across grain boundaries. Acta Mater. 2014; 65:150.

Research supported by the U.S. Department of Energy, Office of Basic energy Sciences, Division of materials Sciences and engineering under Award DE-FG02-08ER46525.

EPJ Web of Conferences **115**, 04002 (2016)
DOI: 10.1051/epjconf/201611504002

MINOS
Centre of Excellence for Nuclear

2nd **Int. Workshop** Irradiation of Nuclear Materials: Flux and Dose Effe
November 4-6, 2015, CEA – INSTN Cadarache, France

Effect of Clear Bands on Intergranular Stresses and IASCC Early Damage

Maxime SAUZAY[1], Mohamed OULD MOUSSA[1], Bob DIAWARA[2], François LEBREAU[2]

[1]CEA-DEN-DMN, Service de Recherches Métallurgiques Appliquées, SRMA (Saclay, France)
[2]CNRS-ENSCP (UMR7045), Laboratoire de Physico-Chimie des Surfaces, Paris-Tech, (Paris, France)

Slip localization is a common feature in post-irradiated metallic poly-crystals undergoing tensile straining. This effect takes place for instance in the form of thin slip bands called channels or clear bands, formed after the local vanishing of irradiation defects induced interactions with gliding dislocations [1-3]. Channel impingement towards grain boundaries (GBs) should induce local stress concentrations along GBs, in the quasi-elastic surrounding matrix. It has been shown extensively that this trigger GB crack initiation [2-3].

Since the fifties, the clear band stress fields have been modeled using the dislocation pile-up theory, which leads to stress singularities similar to the LEFM ones [4]. But such theory does not allow fair predictions of GB fracture, neither in inert or PWR environment [5]. In practice, channel thickness is at least 50 nm depending on material, temperature and loading conditions. As a matter of fact, many slip planes are plastically activated through the channel thickness. Numerous crystalline finite element (FE) computations have been carried out using microstructure inputs varying in broad ranges (slip band aspect ratio and spacing). Slip bands (low critical resolved shear stress) are embedded in a matrix or small aggregates (high CRSS). Microstructure inputs as well as plasticity parameters are evaluated based on TEM observations and dislocation dynamics computation results. High local stress fields are highlighted (Fig. 1a) but they are nevertheless considerably lower than the ones deduced from the pile-up theory (Fig. 1b). Analytical formulae are deduced from the numerous FE results, accounting for channel thickness, grain size, channel density and crystal / GB orientations [6, 7].

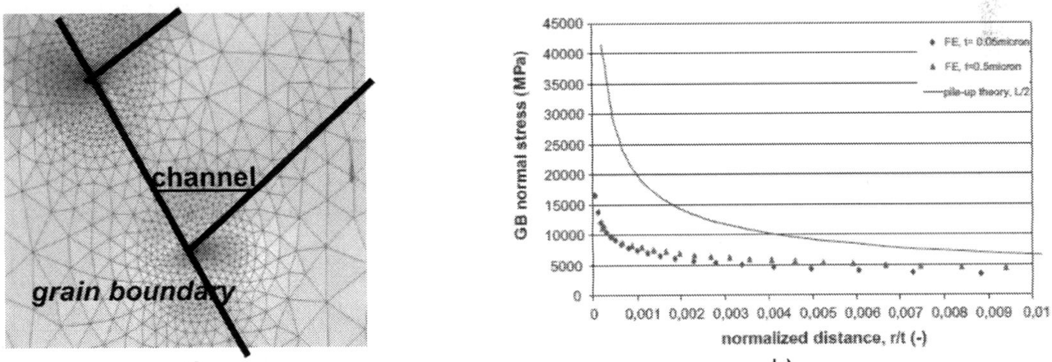

a) b)

Fig. 1: a) local stress concentration at the GB; b) comparison between the GB stress fields computed using the pile-up theory [4] or by the FE method for various clear band thicknesses.

Grain boundary fracture is simulated using a double criterion based on both critical normal stress and fracture energy as deduced from atomistic computations of GB fracture. The critical stress is deduced from the fracture energy using the universal-binding-energy relationship (UBER). In the case of brittle fracture, the fracture energy is defined as the two fresh free surface energy values minus the GB energy. A close-from expression giving the remote tensile stress to GB fracture is then deduced [6,7]. Finally, GB fracture of pre-irradiated SS (dose > 10 dpa) loaded in either inert or pressurized water reactor (PWR) environment is predicted for comparison with the existing experimental results obtained in similar conditions.

In the case of inert environment, free surface and GB energy values are easily found in literature. The predicted remote tensile stress to GB fracture is then equal to the yield stress in agreement with various experimental data (Fig. 2a). In the case of PWR environment, GBs are assumed to be oxidized up to a depth of a few µm due to Cr depletion induced by the strong RIS observed at high irradiation dose [2]. Either literature data have been used (iron oxides [8]) or they have been computed by the DFT method for various GBs and by accounting for hydratation / hydroxylation of the fresh free surfaces in PWR water (Cr_2O_3 oxide). Using the lowest fracture energies corresponding to the weakest GBs, the predicted remote tensile stress to oxidized GB fracture is about one-half of the yield stress (Fig. 2b), once more in fair agreement with many experimental data despite no parameter fitting is required.

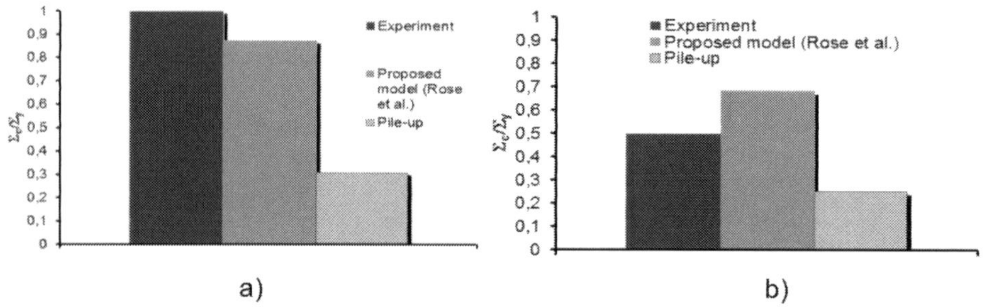

Fig. 2: predicted and measured remote stress to GB crack initiation: a) inert environment [6]; b) PWR environment (Cr_2O_3 oxide).

In relationship with the more complex topic of irradiation induced stress corrosion cracking (IASCC) occurring in-service conditions, the influence of irradiation dose, strain rate and irradiation creep is finally quantitatively discussed. Due to the complex loading history highlighted by structural FE analysis, strain peaks may induce the formation of clear bands or twin bands. The study of the effect of irradiation creep on the relaxation of the induced GB stress fields shows clearly that the relaxation requires many months due to the linear irradiation creep law. Such durations allows GB oxidation mechanisms to occur leading to a weakening of the GB fracture parameters in the vicinity of the free surfaces. This allows the previous scenario of early damage initiation to occur. Finally, the effect cold-working which induces slip localization in the as-received material is discussed. This research work was supported by the European project PERFORM60 (FP7-232612).

References

[1] J. S. Robach, I. M. Robertson, B. D. Wirth, A. Arsenlis, In-situ transmission electron microscopy observations and molecular dynamics simulations of dislocation-defect interactions in ion-irradiated copper. Phil. Mag. 83 (2003) 955.
[2] G.S. Was, Fundamentals of radiation materials science: metals and alloys. Springer (2007).
[3] B Cui, J Kacher, M McMurtrey, GS Was, IM Robertson. Influence of irradiation damage on slip transfer across grain boundaries. Acta Mater. 65 (2014) 15.
[4] A.N. Stroh, A theory of the fracture of metals. Adv. Physics 6 (1957) 418.
[5] P. Evrard, M. Sauzay, Modelling of the effect of dislocation channel on intergranular microcrack nucleation in pre-irradiated austenitic stainless steels during low strain rate tensile loading. J. Nucl. Mater. 405 (2010) 83.
[6] M. Sauzay, K. Vor, Influence of plastic slip localization on grain boundary stress fields and microcrack nucleation. Eng. Fract. Mech. 110 (2013) 330
[7] M. Sauzay, M. Ould Moussa, Prediction of grain boundary stress fields and microcrack initiation induced by slip band impingement, Int. J. Fract. 184 (2013) 215.
[9] De Leew, T. G. Cooper, Geochim. Cosmochim, Surface simulation studies of the hydration of white rust Fe(OH)(2), goethite alpha-FeO(OH) and hematite alpha-Fe_2O_3. Acta 71 (2007) 1655.

EPJ Web of Conferences 115, 04003 (2016)
DOI: 10.1051/epjconf/201611504003

MINOS
Centre of Excellence for Nuclear Materials
2nd Int. Workshop Irradiation of Nuclear Materials: Flux and Dose Effect
November 4-6, 2015, CEA - INSTN Cadarache, France

Evolution under Irradiation of Optimized Austenitic Steel For Gen-IV Reactors. Impact on Fuel Cladding Properties and Performances.

Arnaud COURCELLE[1] Caroline BISOR[1], Emma PIOZIN[1], Mikael KOUNTCHOU[1], Pierre GAVOILLE[1], Marion LE FLEM[2], Jean-Louis SÉRAN[2]

[1]CEA-DEN-DMN, Service d'Etudes des Matériaux Irradiés, SEMI (Saclay, France)
[2]CEA-DEN-DMN, Department of Nuclear Materials (Saclay, France)

An important class of swelling-resistant alloys used in fast-reactor are the cold-worked, titanium-stabilized austenitic stainless steels (15-15Ti AIM1 in France, DIN1.4970 in Germany, optimized D9 in the US and AFIC1 in India). An advanced 15-15Ti alloy was selected for the fuel cladding of the initial core of the fast-reactor French project ASTRID [1]. To extend its qualification at high dose, post-irradiation examinations were recently performed on mixed-oxide fuel pins, irradiated in the Phénix fast-reactor up to 130 dpa and retrieved after Phénix shutdown. As shown in Fig. 1, destructive examinations in hot labs (CEA/LECA-STAR and CEA/LECI facilities) were done to quantify the contribution of irradiation creep and swelling to the total cladding deformation. The results are compared with the past examinations on the same material at lower dose 80 and 110 dpa [2] and confirm the very good swelling resistance of this optimized alloy at 130 dpa.

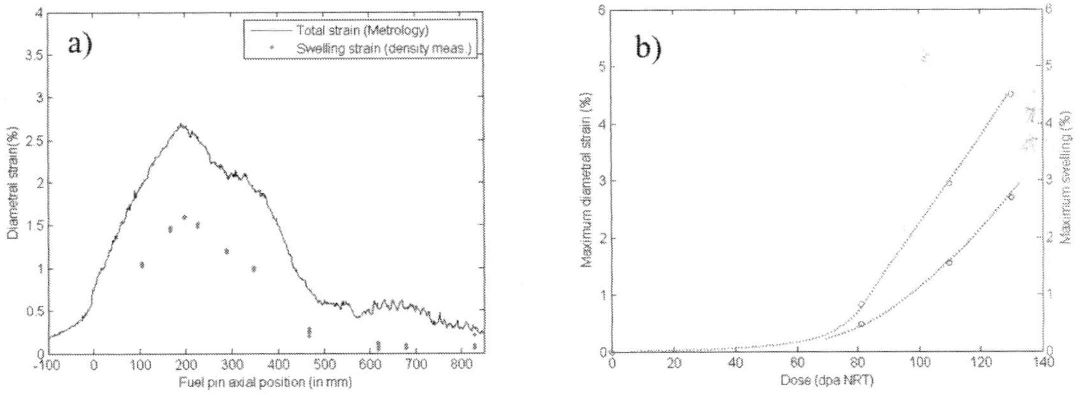

Fig. 1: a) Diametral strain of an irradiated 15-15Ti (optimized) cladding at a maximum dose of 130 dpa. b) Maximum diametral strain and swelling versus the maximum cladding dose.

The evolution of 15-15Ti microstructure under neutron irradiation was investigated by TEM [3]. A shown in Fig. 2, the irradiated microstructure features complex interactions between vacancy clusters, precipitates (fine titanium-carbide and coarser Cr-rich precipitates, G phase) and dislocations network. The simulation of irradiation effects on advanced 15-15Ti using ions irradiation in the JANNUS-Saclay facility will be presented. The comparison between single (2 MeV Fe^{2+}) and dual beams (2 MeV Fe^{2+} - 1 MeV He^{+}) experiments stresses the importance of helium in swelling mechanisms. As in neutron irradiation, heterogeneous nucleation of vacancy clusters is evidenced but marked differences in precipitation mechanisms are observed between neutron and ions irradiation.

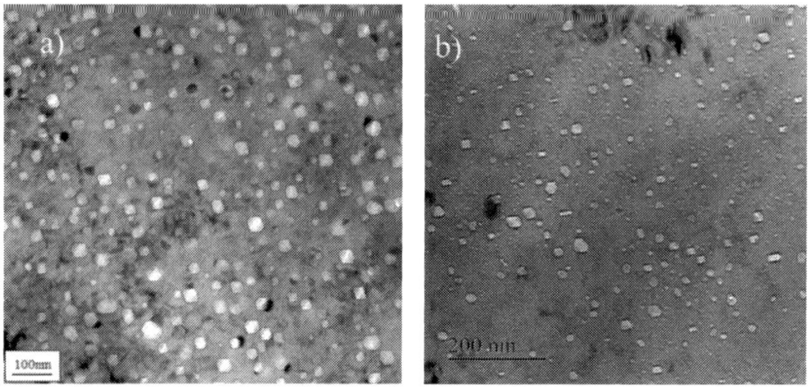

Fig. 2: TEM micrographs of optimized 15-15Ti samples showing cavities microstructure and the interaction between precipitates and voids : a) neutron irradiation 88 dpaNRT 470°C, b) ions irradiation 90 dpaKP 630°C.

The modeling of irradiation evolution in such complex alloy is challenging. The presentation will summarize the current efforts to develop a model capable of predicting temperature, dose, and flux effects on swelling using cluster-dynamics techniques. The modeling of helium effects is also discussed. One of the future areas of work is to include the effects of precipitation (interface effects, helium and point-defects trapping, solid-solution decay etc.).

The Impact of irradiation on tensile and creep properties of 15-15Ti is also briefly discussed with special emphasis on correlations between swelling and embrittlement. At high temperature >600°C, limited data on cold-worked 316Ti and 15-15Ti shows that creep rupture time and ductility are reduced and intergranular failure is enhanced.

References

[1] Le Flem et al. "Status of the French R&D on materials for the core components of ASTRID", International Congress on Advances in Nuclear Power Plants, ICAPP 2014, Charlotte, USA, April 6-9 2014.
[2] A. Maillard et al. "Swelling and Irradiation Creep of Neutron Irradiated 316Ti and 15-15Ti steels", Effects of Radiation in Metals : 16th International Symposium, ASTM STP 1175, 1992.
[3] C. Bisor et al., Personal communication CEA, 2012.
[4] M. Kountchou et al. "Ions Irradiations of Cladding Materials for the Fourth-Generation Nuclear Reactors", Material Science and Technoloy, MS&T 2015, Colombus, USA, October 4-8 2015.

EPJ Web of Conferences **115**, 04004 (2016)
DOI: 10.1051/epjconf/201611504004

MINOS Centre of Excellence for Nuclear Materials
2nd **Int. Workshop** Irradiation of Nuclear Materials: Flux and Dose Effects
November 4-6, 2015, CEA - INSTN Cadarache, France

Multiscale Modeling of Radiation Hardening in RPV and Austenitic Stainless Steels: from the Atomic to the Continuum Scale

Ghiath MONNET[1]

[1]EDF – R&D, Centre Les Renardières, (Moret sur Loing, France)

Given the nanometric size of radiation defects, Radiation Induced Hardening (RIH) is a textbook case of multiscale physical phenomena. On the one hand, pure atomic features that cannot be described by elasticity contribute strongly to the dislocation interaction. Individual interactions – depending on temperature, dislocation velocity, defect nature and size – can be characterized only when these feature are fully identified and quantified. On the other hand, RIH is usually observed at the macroscopic scale, where its technological impact is the deepest. Physically based modeling of RIH must thus integrate materials properties from the atomic level. Since explicit and direct integration is not possible, investigations at intermediate scales associated with appropriate transition methods are thus necessary. Pertinent links are known to be the intragranular and the crystalline aggregate scales. The first one provides constitutive equations controlling plastic flow, including RIH, and the second one delivers the homogenized mechanical behavior.

In this paper, we present recent progress in multiscale modeling of RIH in Reactor Pressure Vessel (RPV) steels and internal steels. We first show that the yield stress prior to irradiation has a non-negligible effect on the RIH, which is found to superpose quadratically with the forest and precipitate hardening. Then, we show how Dislocation Dynamics (DD) simulations can integrate atomistic results and elaborate constitutive equations of plastic behavior. In the following, we present two typical application of our modeling approach.

In internal steels, RIH at doses of some dpa is known to increase tremendously the yield stress. However, the density and the size of the observed radiation defects – formed mostly of dislocation loops – are found (thanks to DD simulations) to provide weak hardening, even when they are considered of extreme strength. The origin of RIH looked difficult to explain. Recently, Atom Probe Tomography (APT) characterization of radiated internal steels have invariably revealed a large density of nanometric solute clusters, which were identified as segregation on small dislocation loops [1].

On the other hand, atomistic simulations have shown that dislocation interactions with dislocation loops leads very often to absorption, as long as the loop is small enough. These experimental observations coupled to atomistic findings inspired a recent investigation [2] in which dislocation interaction with a large number of small dislocation loops were simulated using the DD technique. Since the outcome of individual interactions is invariably absorption, only collinear dislocation loops (i.e. sharing the same Burgers vector as the mobile dislocation) were considered. Interaction with a pure edge dislocation is found to initiate a collective motion of loops in front of the moving dislocation. In contrast, interaction with a pure screw dislocation leads to a large number of collinear interactions strongly pinning the dislocation by forming many helical turns. The stress increment associated with the presence of these loops is depicted on Fig 1a. Hardening is seen to scale with $\mu b\sqrt{DC}$, where μ, b, D and C are respectively the shear modulus, the Burgers vector, the loop size and density. It is clearly seen that the slop of the straight line is close to 0.5, representing the average interaction coefficient. This value falls within the values reported in experiment. Consequently, results suggest that RIH in internal steels can be attributed to the so called "black dots", which are frequently observed in experiments, rather than the faulted (or unfaulted) large dislocation loops that can be easily resolved in Transmission electron microscopy.

The second example, deals with the possible hardening induced by solute clusters, frequently observed in RPV steels. Since solute clusters are directly revealed by APT in all RPV steels and irradiation conditions, they are suspected to be at the origin of RIH in RPV steels. The main difficulty is that it is not possible to use atomistic simulations to compute their resistance, since they are of unknown structure formed of several elements. It is thus necessary to compute the induced hardening as a function of the shear resistance that can be attributed to them. To do so, DD simulations were used to compute hardening induced by a distribution of precipitates of different size, density and shear resistance [3].

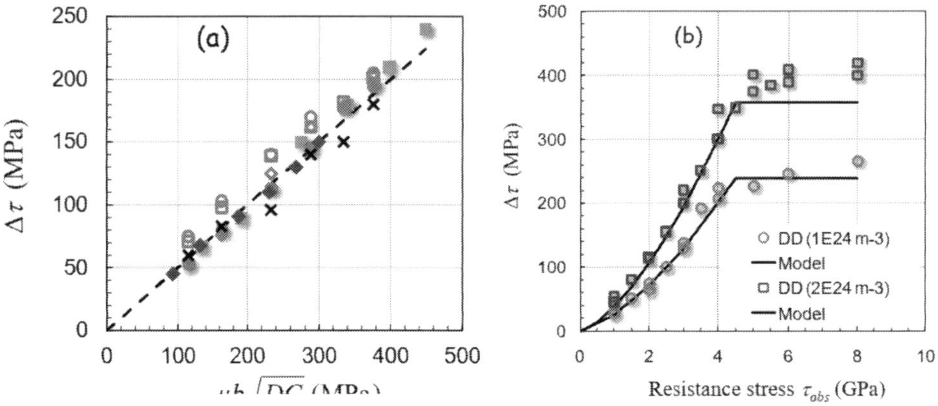

Fig. 1: (a) hardening vs the Orowan stress of collinear dislocation loops [1] and (b) hardening as a function precipitate resistance of 2 nm and two densities [2]. Curves are predictions of the corresponding constitutive equations.

These investigations have allowed establishing a simple expression predicting precipitation hardening as a function of the precipitate size, density and resistance, as confirmed on Fig. 1b. With this constitutive equation it is possible to provide a good approximation of the solute clusters on the yield stress of RPV steels. Comparison with experiment allows to easily identifying the associated shear resistance.

References

[1] Edwards, D. J.; Simonen, E. P. & Bruemmer, S. M. Evolution of fine-scale defects in stainless steels neutron-irradiated at 275 °C Journal of Nuclear Materials, 2003, 317, 13-31.

[2] Monnet, G. New insights into radiation hardening in face-centered cubic alloys Scripta Materialia, 2015, 100, 24-2.

[3] Monnet, G. Multiscale modeling of precipitation hardening: Application to the Fe–Cr alloys Acta Materialia, 2015, 95, 302-311.

EPJ Web of Conferences **115**, 04005 (2016)
DOI: 10.1051/epjconf/201611504005

MINOS Centre of Excellence for Nuclear Materials
2nd *Int. Workshop Irradiation of Nuclear Materials: Flux and Dose Effects*
November 4-6, 2015, CEA - INSTN Cadarache, France

High Burn-up Structure in Nuclear Fuel: Impact on Fuel Behavior

Jean NOIROT[1], Yves PONTILLON[1], Jérôme LAMONTAGNE[2], Isabelle ZACHARIE-AUBRUN[1], Karine HANIFI[1], Philippe BIENVENU[1], Lionel DESGRANGES[3]

[1]CEA-DEN-DEC, Service d'Analyses et de Caractérisation du Comportement des Combustibles, SA3C (Cadarache, France)
[2]CEA-DEN-DEC, Service Fabrication, Elaboration et Reconditionnement des Combustibles, SFER (Cadarache, France)
[3]CEA-DEN-DEC, Service d'Etudes et de Simulation du Comportement des Combustibles, SESC (Cadarache, France)

When UO_2 and $(U,Pu)O_2$ fuels locally reach high burn-up, a major change in the microstructure takes place. The initial grains are replaced by thousands of much smaller grains, fission gases form micrometric bubbles and metallic fission products form precipitates. This occurs typically at the rim of the pellets and in heterogeneous MOX fuel Pu rich agglomerates. The high burn-up at the rim of the pellets is due to a high capture of epithermal neutrons by ^{238}U leading locally to a higher concentration of fissile Pu than in the rest of the pellet. In the heterogeneous MOX fuels, this rim effect is also active, but most of the high burn-up structure (HBS) formation is linked to the high local concentration of fissile Pu in the Pu agglomerates. This Pu distribution leads to sharp borders between HBS and non-HBS areas (Fig.1).

Fig. 1: Sharp limit between HBS and non-HBS areas in a 55 GWd/t_{HM} MOX MIMAS fuel (from [1]).

In these MOX fuels, the HBS forming at various radial positions, and not only at the rim, it was shown that the size of the new grains, of the bubbles and of the precipitates increase with the irradiation local temperatures. Other parameters have been shown to have an influence on the HBS initiation threshold, such as the irradiation density rate, the fuel composition with an effect of the Pu presence, but also of the Gd concentration in poisoned fuels, some of the studied additives, like Cr, and, maybe some of the impurities. However, not all the differences in the UO_2 HBS rim extent measured by different teams on various fuels have been explained [2-3]. The effect of impurities may be the main reason for these differences, but it has not been documented enough yet. It has been shown recently, with examinations of a UO_2 fuel in which ^{235}U was heterogeneously distributed, that a high Pu concentration is not mandatory for HBS formation [4].

Several changes in the fuel behavior occur concomitantly with the HBS formation. An increase of the fission gas release and an increase in the fuel swelling rate are measured [3]. It was shown by indirect and direct approaches that HBS formation was not the main contributor to the increase of fission gas release at high burn-up [1, 5-6]. Indeed, studying the Kr/Xe ratio or some isotopic ratio in the gases collected during rod puncturing and using the differences in the gas productions as a function of the radial position, it was shown that the HBS areas were not the main source of the released gases. SIMS measurements of the local retention confirmed this trend. Nonetheless, the formation of a strong bonding between the fuel pellet periphery and the inner surface of the cladding, with the formation of the inner zirconia layer induces tensile stresses in the fuel during power decrease periods. This leads to radial cracking of the rim and, consequently, to some fission gas release. The amount of gases released by this mechanism never was evaluated. HBS formation does participate to fuel swelling increase, but it is not the only phenomenon involved, other areas showing also gas bubble formation. It must be noted that all porosity observed is not new, a local decrease of the large fabrication pores being measured. The pressure in the HBS bubbles has been evaluated. At 650 K, it is about 78 MPa, and decreases with the increasing burn-up. This corresponds to an average atomic volume of ~150 Å^3, i.e. more than three times the UO_2 trivacancy volume, hence the increase in the swelling [1].

Impact of HBS on the fuel behavior during ramp on high burn-up fuels is unclear. The generalized fuel to cladding contact prior to the ramp is certainly the major parameter. HBS may play a role, but no direct evidence of that was found, though, prior to the ramps, high strain of the HBS can be seen in the large fabrication pores decrease as well as on free surface HBS major swelling. In loss of coolant accident (LOCA) type conditions, out of pile heating tests on fuel sections as well on specially designed experimental discs have shown, like during in-pile LOCA tests, the fragmentation of HBS [7-8]. The areas prone to fragmentation are the HBS areas but also the high precipitation areas [9].
The particular power history of the heterogeneous UO_2, with an increase in the linear powers at high burn-up, led to a generalized fission gas release of the HBS spot bubbles. In addition, a generalized opening of grain boundaries is evidenced by Cs departures and deposits [4].

It could therefore be interesting to design a fuel that would be particularly resistant to HBS formation and to gas precipitation into large bubbles at high temperature.
A large grain UO_2 fabricated by NFD (Japan), with a long oxidizing sintering, was shown to be locally highly resistant both to bubble formation at high temperature and to HBS [10]. Expanding this to the whole fuel is an appealing idea.

References

[1] J. Noirot, L. Desgranges, J. Lamontagne, *Detailed characterisations of high burn-up structures in oxide fuels*, Journ. of Nucl. Mater. 372: 318-339, 2008.

[2] R. Manzel, C.T. Walker, *EPMA and SEM of fuel samples from PWR rods with an average burn-up of around 100 MWd/kg$_{HM}$*, Journ. of Nucl. Mater. 301: 170-182, 2002.

[3] J. Noirot, I. Aubrun, L. Desgranges, K. Hanifi, J. Lamontagne, B. Pasquet, C. Valot, P. Blanpain, H. Cognon, *High Burnup Changes in UO_2 fuels irradiated up to 83 GWd/t in M5 Claddings*, Nuclear Engineering and Technology 41: 155-162, 2009.

[4] J. Noirot, J. Lamontagne, N. Nakae, T. Kitagawa, Y. Kosaka, T. Tverberg, *Heterogeneous UO2 fuel irradiated up to a high burn-up: Investigation of the HBS and of fission product releases*, Journ. of Nucl. Mater. 442: 309-319, 2013.

[5] J. Noirot, L. Desgranges, P. Marimbeau, *Contribution of the rim to the overall fission gas release: what do isotopic analyses reveal.* Fission Gas Behaviour in Water Reactor Fuels, NEA, Ed. NEA/OECD, Cadarache (France), 2000; Vol. NEA #03053, 2002, p 223.

[6] N. Itagaki, K. Ohira, K. Tsuda, G. Fischer, T. Ota, *Fission gas release and pellet microstructure change of high burnup BWR fuel*, Technical Committee meeting on advances in fuel pellet technology for improved performance at high burnup, Tokyo (Japan), IAEA.

[7] Y. Pontillon, M.P. Ferroud-Plattet, D. Parrat, S. Ravel, G. Ducros, C. Struzik, I. Aubrun, G. Eminet, J. Lamontagne, J. Noirot, A. Harrer *Experimental and theoretical investigation of fission gas release from UO2 up to 70 GWd/t under simulated LOCA type conditions: the GASPARD program* LWR Fuel Performance, Orlando, September 19-22, 2004; pp 490-499, Paper 1025.

[8] J. Noirot, Y. Pontillon, S. Yagnik, J.A. Turnbull, T. Tverberg, *Fission gas release behaviour of a 103 GWd/t$_{HM}$ fuel disc during a 1200°C annealing test*, Journ. of Nucl. Mater. 446: 163-171, 2014

[9] J. Noirot, T. Blay, J. Lamontagne, L. Fayette, Y. Pontillon, X. Pujol, *Size and radial origin of fragments formed while heating a 83 GWd/tU PWR fuel up to 1200 °C*. LOCA Workshop, Fuel fragmentation, relocation and dispersal (FFRD), Aix-en-Provence (France), 2015.

[10] J. Noirot, Y. Pontillon, S. Yagnik, J.A. Turnbull, *Post-irradiation examinations and high-temperature tests on undoped large-grain UO2 discs*, Journ. of Nucl. Mater. 462: 77-84, 2015.

EPJ Web of Conferences **115**, 04006 (2016)
DOI: 10.1051/epjconf/201611504006

MINOS
Centre of Excellence for Nuclear Materials

2nd **Int. Workshop** Irradiation of Nuclear Materials: Flux and Dose Effect
November 4-6, 2015, CEA – INSTN Cadarache, France

Influence of irradiation on stainless steel corrosion in PWR primary conditions

Stéphane PERRIN[1], Marie DUMERVAL[1], Loic MARCHETTI[1] Mohamed SENNOUR[2], François JOMARD[3], Sylvain VAUBAILLON[4], Yves WOUTERS[5]

[1]CEA-DEN-DPC, Service de la Corrosion et du Comportement des Matériaux dans leur Environnement, SCCME (Saclay, France)
[2]MINES ParisTech, PSL - Research University, MAT – Centre des matériaux (Evry, France)
[3]CNRS, Groupe d'Etude de la MAtière Condensée (Versailles, France)
[4]CEA, RHF, INSTN, UEPTN (Saclay, France)
[5] SIMaP, Université de Grenoble, (Grenoble, France)

Some internal components of pressurized water reactor (PWR) are made on stainless steel. They are exposed to:

- a corrosive environment which is the primary medium that is to say water at high temperature (290°C-330°C) and high pressure (155 bars) with boric acid (H_3BO_3) , lithium hydroxide (LiOH) and dissolved hydrogen,
- a mechanical stress,
- an irradiation process due to neutron flux which can cause up to 1 displacement per atom (dpa) by year.

In these conditions, these materials can be damaged by a process of IASCC (Irradiation Assisted Stress Corrosion Cracking). This degradation process has been observed since the eighties with the formation of cracks on baffle bolts under operating conditions. Oxide film rupture is the first step in this corrosion process and thereby the properties of the oxide formed on stainless steels are key parameters relative to SCC initiation. The morphology and the structure of the oxide formed on stainless steel in PWR primary medium have been detailed by several authors [1, 2, 3]. The oxide layer in divided in two parts: an internal and continuous layer rich in chromium and an external discontinuous layer composed of crystallites rich in iron. Moreover a nickel enrichment has been observed in the alloy under the oxide close to the alloy/oxide interface. In this alloy layer with a higher composition in nickel, a trapping of hydrogen has also been observed [4]. The aim of this work is to study the influence of defects created in the alloy by irradiation on the oxide properties (thickness, morphology, composition) and on oxygen transport in this oxide scale during exposition of stainless steel in PWR primary medium. In order to simulate the defects due to irradiation, ionic implantation have been performed. So samples of austenitic stainless steel (316L) have been implanted with Xenon ions at energy of 240 keV. Transmission Electron Microscopy (TEM) observations indicate an affected depth of about 80 nm with the formation of small cavities. The implanted samples have been oxidized at 325°C in hydrogenated primary water which simulates the primary medium in different devices, static autoclave and corrosion loop during different durations. The oxide film properties have been investigated by a set of techniques (SEM, TEM, SIMS) .As shown on Fig. 1, which presents implanted sample before and after 600 hours of oxidation, it has been observed a correlation between the affected area created by implantation and the morphology of the oxide formed after corrosion test. The composition of the internal layer is also linked to the defects created by implantation.

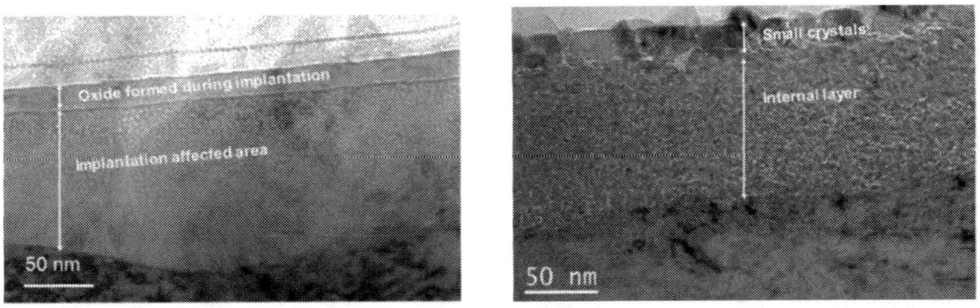

Fig. 1: TEM images on implanted 316L samples before (a) and after (b) 600 h of oxidation

Moreover the study of oxidation kinetic evaluated by ion beam analysis (NRA and RBS) have shown that the growth rate of the internal oxide layer follows a logarithmic-type law: after a 24 h exposure, the internal oxide reaches a thickness equivalent to the implanted affected depth which does not change much for longer exposition duration confirming the influence of the defect on the oxidation process.

Oxygen transport in the oxide layer has also been studied by two-stages corrosion experiments, a first stage in natural primary water (during 600h) followed by a second stage with water enriched in ^{18}O (during 16h). Comparison between TEM observation after implantation, EDX analysis after corrosion and SIMS analysis after these two corrosion sequences (Fig. 2) enable to link the distribution of defect created by implantation with the oxygen transport through the internal oxide layer formed during exposition in primary medium at 325°C.

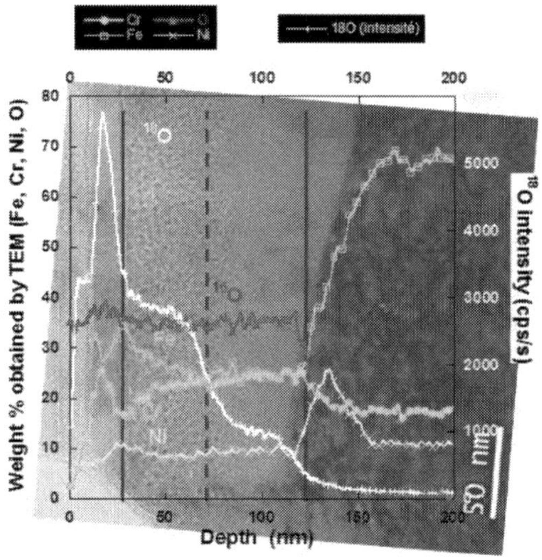

Fig 2: Superposition of TEM image showing the microstructure samples after implantation, concentration profiles of ^{18}O obtained by SIMS after two corrosion stages and O, Fe, Cr and Ni EDX concentration profiles obtained after 600 h of exposition to primary medium.

As a conclusion, this study revealed an effect of the defect distribution (generated by prior Xe-implantation) on the oxide formed on of 316L in primary medium, on the oxidation kinetic and also on the oxygen transport properties in the oxide layer.

References

[1] Terachi, T. Yamada, T. Miyamoto, K. Arioka, K. Fukuyai, Corrosion Behavior of Stainless Steels in Simulated PWR Primary Water -Effect of Chromium Content in Alloys and Dissolved Hydrogen, Journal of Nuclear Science and Technology, 45 (2008) 975

[2] B. Stellwag, The mechanism of oxide film formation on austenitic stainless steels in high temperature water, Corrosion Science, 40, (1998) 337

[3] S. Perrin, L. Marchetti, C. Duhamel, M. Sennour and F. Jomard, Influence of irradiation on the oxide film formed on 316L stainless steel in PWR primary water,.Oxidation of Metals, 80 (2013) 623

[4] M. Dumerval, S. Perrin, L. Marchetti, M. Tabarant, F. Jomard, Y. Wouters, Hydrogen absorption associated with the corrosion mechanism of 316L stainless steels in primary medium of Pressurized Water Reactor (PWR), Corrosion Science, 85, (2014) 251

EDP Sciences – Web of Conferences
17, Avenue du Hoggar
Parc d'Activité de Courtabouf
BP 112
F-91944 Les Ulis Cedex A
France

ISBN 978-1-5108-2209-2

CURRAN ASSOCIATES INC.
proceedings.com